Tensile and Microstructural Characterization of Friction Stir Welded AA2219-T87

By
Manoj

TABLE OF CONTENTS

Page No.

CHAPTER 1

INTRODUCTION

1.1 Foundation and Purpose of the Investigation

The various current applications and production requirements of complex components using lightweight multi-materials such as advanced high strength steel (AHSS), aluminium, and magnesium alloys require proper welding methods. When welding all aluminium alloys, significant degradation in hardness and tensile strength can be expected in the HAZ. In many cases, as much as 50% reduction in properties will occur. Softening occurs in the HAZ of cold worked aluminium alloys due to recrystallization and grain growth. I. Baker, (2000) There is no way to recover this loss in properties unless the material is cold worked again, which is rarely a practical solution. As a result, cold worked aluminium alloys are often considered unweldable, not because they cannot be welded but because of the drastic loss in unrecoverable properties. Manwatkar, S. K., et al (2019) have studied the cracking in aluminium alloy AA2219 TIG welded joints and attributed the problem to improper fixturing, large heat input along with large, grained microstructure. The deleterious effect of coarse grain size can be mitigated by proper weld parameter selection with proper fixturing to avoid.

Bush, R., Feier, I., et al (2024) found that the weld HAZ softening that occurs when welding precipitation-strengthened alloys is due to over aging and/or dissolution of the strengthening precipitates that were formed during the precipitation hardening treatment. When welding these alloys, however, there is an opportunity to recover some or all the properties through post weld heat treatments. One approach is to solutionize the entire weldment to completely dissolve the precipitates, followed by an aging treatment. This approach is effective, but time consuming, costly, and in some cases not practical. Another approach is to weld in the “T4” condition worked Rana, A., Deshwal, S., et al. (2024). This condition includes a “natural” aging treatment that does not fully strengthen the alloy. When an alloy in the “T4” condition is welded, the amount of precipitate over aging is reduced, but not eliminated. Nath, S., Konkati, C., et al. (2024) The weld can then be aged to achieve near base metal strength levels. Aluminium is well known for its propensity to form weld metal porosity. This is due to the fact that molten aluminium can dissolve considerable amounts of hydrogen, especially at higher temperatures. Once aluminium melts (660°C/1220°F), its solubility for hydrogen increases dramatically. The solubility further increases at a rapid rate upon further heating. During welding, as the hot molten aluminium in the weld metal begins to cool, its solubility for hydrogen drops rapidly. This causes the hydrogen to come out of solution and form gas bubbles. Upon further cooling and subsequent solidification, the bubbles become trapped, resulting in porosity. The best approach during welding is to keep heat input very low to minimize the loss in strength, or to weld in the fully annealed condition.

1.2 About Aluminum Alloys

Since the early 1990s, the use of aluminum has grown significantly driven by advancements in technology and scientific research. Tabereaux, A. T., & Peterson, R. D. et al (2024) These developments have expanded the applications of existing materials and accelerated the search for high-performance materials that can meet the demands of modern industries. Researchers have increasingly shifted towards enhancing the properties of known materials and discovering alternatives with superior characteristics. This ongoing effort is especially important in aviation, space exploration, gas turbines, automotive, and maritime industries, where lightweight, durable, and high-performance materials capable of withstanding extreme conditions are essential. Aluminum and its alloys have gained prominence in aviation due to their low density, high specific strength, and excellent oxidation resistance, making them ideal for long-term, high-demand applications Shen, A., & Zhang, J. et al. (2024). The material's flexibility in production, ease of shaping, high corrosion resistance, and lightweight mechanical properties have led to its extensive use across various industries. Additionally, aluminum's ability to be recycled with only 5% of the energy required for primary production has strengthened its role in sustainable engineering practices. However, the inherent softness of pure aluminum, with a tensile strength of approximately 10 MPa, limits its industrial applications. To overcome this, alloying, particle reinforcement, plastic deformation, and grain size refinement enhance its strength.

Today, aluminum alloys rank second only to steel as the most widely used metallic engineering materials. Their unique properties which include high thermal conductivity, non-flammability, full recyclability, and excellent weldability make them indispensable in various industrial applications. This thesis explores the evolving role of aluminum alloys in modern engineering focusing on their enhanced properties, applications, and the methods used to improve their performance.

1.3 Effects of Alloying Elements and Impurities on Aluminium

The addition of specific elements to pure aluminium which can constitute up to 10% of the alloy composition greatly enhances its properties and usefulness. The major alloying additions used with aluminium are copper, manganese, silicon, magnesium, and zinc. Impurity elements are also present, but their total percentage is usually less than 0.15% in aluminium alloys.

1.3.1 Copper

Aluminium-copper alloys containing 2 to 10% Cu, generally with other additions, form important families of alloys. Both wrought and cast aluminium-copper alloys respond to solution heat treatment and subsequent aging with an increase in strength and hardness and a decrease in elongation. The strengthening is maximum between 4 and 6% Cu, depending upon the influence of other constituents present. Most commercial alloys contain other alloying elements.

1.3.2 Manganese

Manganese has a very limited solid solubility in aluminium in the presence of normal impurities and its concentration normally ranges from 5 to 50 ppm. It decreases resistivity. Manganese increases strength either in solid solution or as a finely precipitated intermetallic phase. It has no adverse effect on corrosion resistance. As an addition, it is used to increase strength and to control the grain structure. The effect of manganese is to increase the recrystallization temperature and to promote the formation of fibrous structure upon hot working. As a dispersed precipitate it is effective in slowing recovery and in preventing grain growth. The manganese precipitate increases the quench sensitivity of heat-treatable alloys. Manganese is the main alloying addition of the 3xxx series of alloys, in which it is added alone or with magnesium. This series of alloys is used in large tonnages for beverage containers and general utility sheet.

1.3.3 Magnesium

Magnesium is the major alloying element in the 5xxx series of alloys. Its maximum solid solubility in aluminium is 17.4%, but the magnesium content in current wrought alloys does not exceed 5.5%. The addition of magnesium markedly increases the strength of aluminium without unduly decreasing the ductility. Corrosion resistance and weldability are good. Magnesium precipitates preferentially at grain boundaries as a highly anodic phase (Mg_5Al_3 or Mg_5Al_8), which produces susceptibility to intergranular cracking and to stress corrosion. Wrought alloys containing up to 5% Mg, properly fabricated, are stable under normal usage.

1.3.4 Nickel

The solid solubility of nickel in aluminium does not exceed 0.04%. Over this amount, it is present as an insoluble intermetallic, usually in combination with iron. Nickel (up to 2%) increases the strength of high-purity aluminium but reduces ductility. Nickel is added to aluminium-copper and to aluminium-silicon alloys to improve hardness and strength at elevated temperatures and to reduce the coefficient of expansion. Nickel promotes pitting corrosion in dilute alloys such as 1100. It is limited in alloys for atomic reactor use, due to its high neutron absorption.

1.3.5 Silicon

After iron, Silicon has the highest impurity level in electrolytic commercial aluminium (0.01 to 0.15%). In wrought alloys, silicon is used with magnesium at levels up to 1.5% to produce Mg2Si in the 6xxx series of heat-treatable alloys. High-purity aluminium-silicon alloys are hot short up to 3% Si, the most critical range being 0.17 to 0.8% Si, but additions of silicon (0.5 to 4.0%) reduce the cracking tendency of aluminium-copper-magnesium alloys. Small amounts of magnesium added to any silicon-containing alloy will render it heat-treatable.

1.3.6 Indium

Small amounts of 0.05 to 0.2% Indium have a marked influence on the age hardening of aluminium-copper alloys, particularly at low copper contents (2 to 3% Cu). In this respect, indium acts very much like cadmium in that it reduces room-temperature aging but increases artificial aging. The addition of magnesium decreases the effect of indium.

1.3.7 Tin

Tin is used as an alloying addition to aluminium from concentrations of 0.03 to several percent in wrought alloys, small amounts of tin (0.05%) greatly increase the response of aluminium-copper alloys to artificial aging following a solution heat treatment. The result is an increase in strength and an improvement in corrosion resistance. Higher concentrations of tin cause hot cracking in aluminium-copper alloys.

1.3.8 Molybdenum

Molybdenum is a very low level (0.1 to 1.0 ppm) impurity in aluminium. It is used at a concentration of 0.3% as a grain refiner, because the aluminium end of the equilibrium diagram is peritectic, and also as a modifier for the iron constituents, but it is not in current use for these purposes.

1.3.9 Lithium

The impurity level of lithium is of the order of a few ppm, but at a level of less than 5 ppm it can promote the discoloration (blue corrosion) of aluminium foil under humid conditions. Traces of lithium greatly increase the oxidation rate of molten aluminium and alter the surface characteristics of wrought products. Binary aluminium-lithium alloys age hardens but are not used commercially. Present interest is on the aluminium copper-magnesium-lithium alloys, which can be heat-treated to strengths comparable to those of 7xxx alloys. In addition, the density is decreased, and the modulus is increased.

1.3.10 Zirconium

Zirconium is added to make aluminium alloys a stronger, more corrosion-resistant and thermal-resistant stable alloy and if the alloy is exposed to high temperatures it will return to its original condition. More specifically, adding Zirconium to aluminium, inhibits recrystallization and allows grain structure to be controlled at elevated temperatures. Zirconium has therefore been instrumental in the development of advanced aluminium overhead conductors that can operate at high temperatures. Aluminium zirconium alloys are in high demand for applications in high-performance industries due to their ability to withstand demanding performance requirements.

1.3.11 Copper-Magnesium

The main benefit of adding magnesium to aluminium-copper alloys is the increased strength possible following solution heat treatment and quenching. In wrought material of certain alloys of this type, an increase in strength accompanied by high ductility occurs on aging at room temperature. On artificial aging, a further increase in strength, especially in yield strength, can be obtained, but at a substantial sacrifice in tensile elongation. On both cast and wrought aluminium-copper alloys, as little as about 0.5% Mg is effective in changing aging characteristics. In wrought products, the effect of magnesium additions on strength can be maximized in artificially aged materials by cold working prior to aging

1.3.12 Copper and Minor Additions

In the wrought form, an alloy family of interest is the one containing small amounts of several metals known to raise the recrystallization temperature of aluminium and its alloys, specifically manganese, titanium, vanadium, or zirconium. An alloy of this nature retains its properties well at elevated temperatures, fabricates readily, and has good casting and welding characteristics. In general, tensile strength increases with separate or simultaneous increases in magnesium and manganese, and the yield strength also increases, but to a lesser extent. Further increases in tensile and particularly yield strength occur on cold working after heat treatment. (J.R. Davis 1993)

1.4 Aluminium Alloy Categories.

Aluminium alloys are broadly categorized into two major classes: wrought alloys and cast alloys. Within each class, a further distinction is made based on the primary mechanism of property enhancement. Many aluminium alloys exhibit phase solubility behaviour and respond to thermal treatments, such as solution heat treatment, quenching, and precipitation (or age) hardening. Alloys that undergo such treatments, whether cast or wrought, are classified as heat-treatable alloys.

In contrast, a significant number of wrought aluminium alloys develop their properties through work hardening, which involves mechanical deformation, typically followed by controlled annealing processes. These alloys do not rely on thermal treatment but instead gain strength and other properties through mechanical reduction. A further differentiation for each category is based on the primary mechanism of property development. Many alloys respond to thermal treatment based on phase solubilities. These treatments include solution heat treatment, quenching, and precipitation, or age, hardening. For either casting or wrought alloys, such alloys are described as heat treatable. A large number of other wrought compositions rely instead on work hardening through mechanical reduction, usually in combination with various annealing procedures for property development. These alloys are referred to as work hardening. (Mondolfo, L. F. (1976), Hatch, J. E. (1984), Polmear, I. J. (1995), Davis, J. R. (1993)

The American Aluminium Association system had defined wrought aluminium alloys into eight classes of families for simplification. Individual alloys are designated using a four-digit number, with the first digit representing the basic class. These classes are distinguished by the primary alloying elements and how they are strengthened either strain hardening by cold working or a precipitation- strengthening using heat treatment

Other digits in the designation have no special significance. As the table shows, the 1xxx, 3xxx, 4xxx, and 5xxx alloys are all strengthened by cold work, whereas the 2xxx, 6xxx, 7xxx, and 8xxx alloys are all strengthened by a precipitation heat treatment. It is important to recognize the strengthening method for the alloy prior to welding. For both cases after welding, there will be considerable degradation of the tensile strength, but it is possible to recover such losses in those alloys that are strengthened by heat treatment. The four-digit alloy code is often followed by a heat treatment, processing, or temper designation code. This code provides some detail regarding how the particular alloy is strengthened. For example, the "T6" in 6061-T6 refers to a solution heat treatment followed by a precipitation-strengthening aging treatment, for wrought alloys a four-digit system is used to produce a list of wrought composition families as follows.

1.4.1 1xxx Series

Aluminium with a purity of 99.00% or higher is widely utilized in applications, particularly within the electrical and chemical industries. These high-purity aluminium grades are noted for their exceptional corrosion resistance, superior thermal and electrical conductivity, low mechanical strength, and excellent formability. The mechanical properties can be moderately enhanced through strain hardening. Iron and silicon are typically the primary impurities in these alloys. Due to their conductive properties, the 1xxx series aluminium alloys are extensively employed as electrical conductors, including in high-voltage transmission lines.

1.4.2 2xxx Series.

The 2xxx series aluminium alloys have copper as the principal alloying element, with magnesium often added as a secondary element. These alloys require solution heat treatment to achieve optimal mechanical properties. In the heat-treated condition, their strength is comparable to, and sometimes exceeds, that of low-carbon steel. Precipitation heat treatment (aging) is often employed to enhance yield strength, though this treatment typically reduces elongation while having a lesser effect on tensile strength.

While the 2xxx series alloys offer high strength, they generally exhibit inferior corrosion resistance compared to other aluminium alloys. Under certain environmental conditions, these alloys can be susceptible to intergranular corrosion. As a result, when used in sheet form, they are often clad with high-purity aluminium, a 6xxx series magnesium-silicon alloy, or an alloy containing around 1% zinc to improve corrosion resistance through galvanic protection. The cladding, typically 2-5% of the total thickness on each side, significantly enhances the overall corrosion resistance. These alloys are especially suitable for applications requiring high strength-to-weight ratios. Common applications include truck and aircraft wheels, truck suspension components, aircraft fuselage and wing skins, and structural parts, especially in conditions where strength is required at temperatures up to 150°C (300°F).

1.4.3 3xxx series

In the 3xxx series of aluminium alloys, manganese is the primary alloying element. These alloys are generally non-heat-treatable, yet they exhibit approximately 20% greater strength compared to the 1xxx series. The effectiveness of manganese as an alloying element is limited to a concentration of up to about 1.5%, which restricts its usage in significant quantities across different alloys. Despite this limitation, alloys such as 3003 are widely utilized due to their moderate strength, excellent workability, and versatility for general-purpose applications.

1.4.4 4xxx series

The 4xxx series aluminium alloys are primarily alloyed with silicon, which can be added in substantial amounts (up to 12%) to significantly lower the melting range without inducing brittleness. Due to this characteristic, aluminium-silicon alloys are commonly used in welding wire and brazing applications where a lower melting point than that of the base metal is required. While most alloys in this series are non-heat-treatable, when they are used in welding heat-treatable alloys, they may absorb some of the alloying elements from the base metal, allowing for limited responsiveness to heat treatment.

Alloys in this series that contain significant amounts of silicon tend to develop a dark grey to charcoal appearance when anodized, making them desirable for architectural applications. Alloy 4032 is notable for its low coefficient of thermal expansion and high wear resistance, making it well-suited for the production of forged engine pistons.

1.4.5 5xxx series

The major alloying element in the 5xxx series aluminium alloys is magnesium. When used as the primary alloying element, either alone or in combination with manganese, magnesium produces a moderate to high-strength alloy that is work-hardenable. Magnesium is significantly more effective as a strengthening agent than manganese, with approximately 0.8% magnesium providing the same strengthening effect as 1.25% manganese. Additionally, magnesium can be added in much higher concentrations. Alloys in this series exhibit excellent welding characteristics and provide outstanding corrosion resistance, particularly in marine environments. However, there are limitations regarding the amount of cold work and the safe operating temperatures for alloys with higher magnesium content (above approximately 3.5%). For operating temperatures exceeding 65°C (150°F), alloys with higher magnesium content can become susceptible to stress-corrosion cracking, so caution is advised in such conditions.

1.4.6 6xxx series

The 6xxx series alloys are composed of silicon and magnesium in ratios that facilitate the formation of magnesium silicide (Mg_2Si), making them amenable to heat treatment. While these alloys generally exhibit lower strength compared to most 2xxx and 7xxx series alloys, they offer a favorable balance of medium strength with excellent formability, weldability, machinability, and corrosion resistance. Alloys within this heat-treatable series can be processed in the T4 temper, where they undergo solution heat treatment without subsequent precipitation hardening. After forming, they can be further strengthened to achieve full T6 properties through precipitation heat treatment. Figure 3 illustrates the relationships among several commonly utilized 6xxx series alloys.

1.4.7 7xxx series

7xxx series Zinc, in concentrations ranging from 1% to 8%, serves as the primary alloying element in 7xxx series aluminium alloys. When combined with magnesium in smaller amounts, these alloys become heat-treatable, exhibiting strength levels from moderate to very high. Small additions of elements such as copper and chromium are commonly included to enhance specific properties, and the incorporation of trace amounts of scandium can further improve performance. These alloys are widely utilized in applications requiring high strength, such as airframe structures and mobile equipment. However, higher-strength variants of the 7xxx series are more susceptible to stress corrosion cracking, necessitating the use of a slightly overaged temper to achieve an optimal balance between strength, corrosion resistance, and fracture toughness.

1.4.8 8xxx series

Alloys encompass a diverse range of chemical compositions designed to achieve specific material properties. For instance, enhanced high-temperature performance can be attained through dispersion-strengthened aluminium alloys such as Al-Fe-Ce (e.g., 8019) or Al-Fe-V-Si (e.g., 8009), typically produced via powder metallurgy processes. Moreover, alloys containing lithium, such as the 8090 series, offer reduced density and increased stiffness. The 8090 alloy, being precipitation hardenable, has replaced certain medium- to high-strength aluminum alloys (e.g., 2xxx and 7xxx series) in aerospace and aircraft applications, including helicopter components, due to its superior properties.

Table 1.1 Classification Wrought Aluminium alloys based on their strengthening mechanism

Alloy system	Aluminum series
Work-hardening alloys	
Pure Al	1xxx
Al-Mn	3xxx
Al-Si	4xxx
Al-Mg	5xxx
Al-Fe	8xxx
Al-Fe-Ni	8xxx
Precipitation-hardening alloys	
Al-Cu	2xxx
Al-Cu-Mg	2xxx
Al-Cu-Li	2xxx
Al-Mg-Si	6xxx
Al-Zn	7xxx
Al-Zn-Mg	7xxx
Al-Zn-Mg-Cu	7xxx
Al-Li-Cu-Mg	8xxx

1.4.9 AA2219-T87 alloy

The 2000-series (Duralumin) aluminium alloys are commonly used in aerospace structures due to their strength-to-weight ratio and relatively high fracture toughness. The aluminium alloy AA2219 -T87 is an age hard enable alloy widely used in the construction of liquid cryogenic rockets. This alloy comprises a large portion of NASA's SLS rocket, as well as previous launch vehicles including Saturn V and Space Shuttle. (Srinivasa Rao et.al.). The alloy is characterized by a unique combination of properties: weldability, high strength and plasticity at a sufficiently low specific gravity, high strength to weight ratio, good fracture toughness and superior oxidation and corrosion resistance as well as significant cryogenic resistance.

High-strength aluminium alloy materials have been increasingly employed in welding of large structures in aerospace, such as launch vehicle, space shuttles and spaceships fuel tanks. The welded structures usually experience a complex internal/external pressure and structure torque during the service which requires the high-standard welds. The conventional fusion welding technique is no longer competitive to fulfil the requirement of rocket fuel tanks with high reliability and high efficiency production. Therefore, aerospace and automotive manufacturers are continuously looking for effective methods to achieve this goal. One of the popular methods to increase the fuel efficiency is to use the auto-body parts made of stronger materials, which will help in down-gauging, leading to a decrease in the vehicle weight [2,3]. Advanced high-strength aluminium alloy is a family of such materials and lightweight, which can be used to replace the conventional heavy materials used in aerospace and auto-body structures. The structural applications of aluminium alloys unavoidably involve welding and joining. One of the most important concerns in fusion welding of the aluminium alloys is softening of the weld metal compared to the wrought and heat-treated base metal which is caused premature failure during the service. This softening is attributed to the decrease in the dislocation density and dissolution or growth of the strengthening precipitates in non-heat treatable and heat treatable aluminium alloys, respectively. Furthermore, this behaviour is directly affected by the heat-input during welding. In fact, the more is the welding heat-input, the higher is the weld metal softening

Al 2219 is a binary Al–Cu alloy which precipitation aging follows the sequence: $\alpha \rightarrow \alpha$ + GP zones$\rightarrow \alpha + \theta'' \rightarrow \alpha + \theta' \rightarrow \alpha + \theta$. The aluminium solid solution is indicated by α, the metastable phases are indicated by θ', θ'' and the stable precipitates by θ. which were formed in the alloy during the aging bellow 200°C and 350°C respectively. The θ particles were formed at larger size particles at high temperature region, due to which, the θ particles were visible at lower magnification. GP zones, θ'' and θ', are metastable phases, but θ is the equilibrium phase in the binary system (J. M. Silcock (1959). The figure 1.1 shows Aluminium –Copper phase diagram

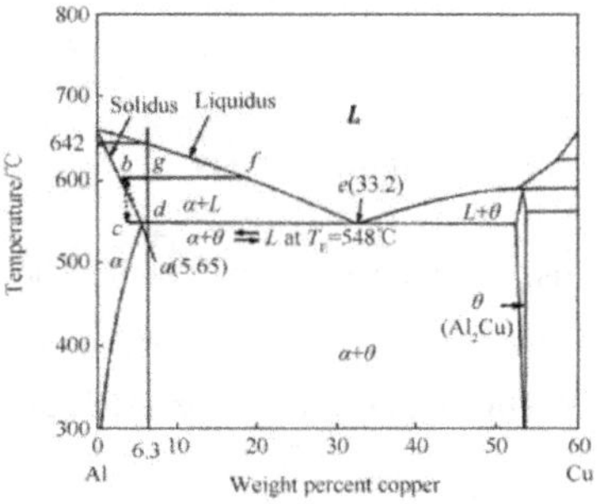

Fig 1.1 Aluminium –Copper phase diagram (J. M. Silcock)

The rate of formation of G.P. zones in Al-cu alloys are dependent on the method of quenching and is much lower after a reversion treatment (Toman, K. (1955), Guinier 1952) (i.e. a short anneal at about 200 °C). This has recently been attributed to variations in the concentration of quenched-in vacancies. The rate of formation of the metastable precipitate θ′ is dependent on dislocations present in the alloy rather than an excess mobile vacancy concentration. Traces of indium (or cadmium) form even more efficient nuclei for & than dislocations but the effect of the indium (or cadmium) is destroyed by fast quenching or cold work. It is known that both cold work and rapid quenching accelerate the rate of formation of θ′ and that room temperature ageing appears to reduce the effect of cold working. (J. M. Silcock ,1959)

To achieve a high strength, precipitate strengthened alloy, it is advantageous to maximize the number of small precipitates and encourage particle shearing. As the material is over-aged and particles increase in size and spacing, the strength will decrease (Porter, 1981).

The alloy is very sensitive to thermal welding cycle; therefore, during melting and crystallization of seams, the metal strength in heat-affected zone decreases due to the formation of heterogeneous structure during segregation of alloying elements and impurities along crystallites boundaries and grains. Seam microstructure is determined by temperature gradient on distribution surface of solid solution, liquid metal and phases, the value of crystallization rate, as well as the nature of distribution of alloying elements in the volume of the seam metal. Due to the inhomogeneous microstructure and mechanical properties in each typical region of the joint, the tensile strength of the joint is always severely reduced (L. I. Nyrkova et al (2023)

1.5 Thermal treatments of aluminium alloy

Aluminium which is thermally treated with or without supplementary strain-hardening is designated with a letter symbol "T". The T is followed by a digit or digits which designate the specific thermal treatment. Temper designations for aluminium alloys are as follows:

T1 - cooled form from an elevated temperature shaping process and naturally aged to a substantially stable condition.

T2 - annealed. This is used to cast products.

T3 - solution heat treated and then cold worked.

T4 - solution heat treated and naturally aged to a substantially stable condition. This is applicable for products which are not cold worked after solution heat treatment or in which effect of cold work in flattering or straightening.

T5 - cooled from elevated temperature shaping process and then artificially aged.

T6 - solution heat treated and then artificially aged.

T7 - solution heat treated and over-aged(stabilized).

T8 - solution heat treated, cold worked and then artificially aged. This is applicable for products which are cold worked to improve strength.

T9 - solution heat treated, artificially aged and then cold worked. This is applicable for products which are cold worked to improve strength.

T10 - cooled form from an elevated temperature shaping process artificially aged and then cold worked. This is applicable for products which are artificially aged after cooled form an elevated temperature shaping process, such as casting or extraction and then cold worked to further improve strength.

T31- Solution heat treated and Strech

T37- Solution heat treated and 8% Cold work

T62-T4 Aged 36 hours at 190^0C

T81- T31Aged 18 hours at 177^0C

T87- T37 Aged 24 hours at 163^0C

1.6 Welding

Welding is a fabrication process that involves joining two or more metal pieces by melting their boundary surfaces, typically through the application of heat, pressure, or both. A filler material is sometimes added to form a strong metallic bond between the joined pieces, resulting in a solid cohesive structure.

1.7 Categorization of Welding Techniques

Welding processes can be broadly classified based on the mechanisms of material joining, specifically temperature and pressure. These processes are generally divided into two main categories: Plastic or Pressure Welding, where the materials are joined through the application of pressure without reaching the melting point, and Fusion or Non-Pressure Welding, where the materials are melted and fused without the need for applied pressure. Each category encompasses various techniques (figure 1.2) tailored to specific industrial applications, relying on different heat sources and joining methods to achieve optimal results.

1.7.1 Plastic or Pressure Welding

Pressure welding involves heating the metals until they reach a plastic state and applying external pressure to achieve bonding. This process, which occurs at temperatures significantly below the melting point of the base metals and without the use of filler material, is characterized by the coalescence of materials through a combination of heat, time, and pressure. As the base metals do not melt, the materials' original mechanical and chemical properties are largely preserved. Solid-state welding, one of the earliest welding techniques, is a prime example of this process. The below figure 1.2 shows the classification of welding processes based on their energy

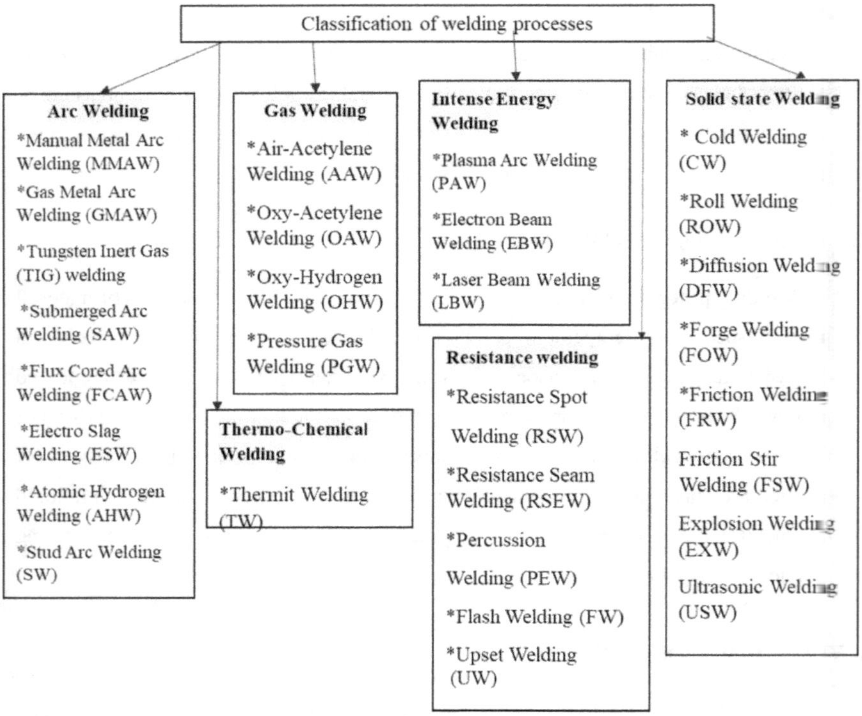

Fig 1.2 Classification of welding processes based on their energy

1.7.2 Fusion or Non-Pressure Welding

In fusion welding, the materials at the joint are heated until they melt, and upon cooling, they solidify to form a strong bond. This method relies solely on heat with no external forces required to join the materials. A filler material is typically added to ensure the joint is filled and secured. Key elements in all fusion welding processes include sufficient heat to melt the materials, shielding to protect the molten weld from exposure to the atmosphere, and a filler material to complete the joint. The process creates a solid and lasting bond by focusing on the melting and solidification of the materials without needing mechanical pressure or force during welding.

1.8 Solid-State Welding

When pressure is applied externally instead of applying heat, then it is termed as solid-state welding. However, in this process heat may produce internally that can elevate the temperature of base materials (but base materials don't fuse as temperature remains lower than melting temperature). Here usually filler materials are not applied (autogenous welding). Resistance and diffusion-based welding processes usually come under this category. For example: FSW

1.9 Conventional Welding methods for Aluminium alloys

A few welding methods are available based on the variability associated with entities joined, corresponding joining mechanism and source of energy used. Aluminium's high thermal conductivity also presents challenges. It dissipates heat quickly, requiring more energy input at the start of welding to achieve proper melting and penetration. As the process continues, heat accumulates in the base metal, necessitating adjustments in current to avoid overheating and distortion.

Conventional methods used for aluminium alloy welding in automotive industry are resistance spot welding (RSW) and self-piercing riveting (SPR) Amongst numerous welding techniques that are generally employed by various industries in welding Aluminium based alloys especially those precipitations strengthened, and solid solution strengthened are GMAW, GTAW arc welding, electron beam and laser welding techniques are the main categories of welding processes. The difficulties associated with the fusion welding of high strength aluminium alloys are the high heat-input associated softening with strength loss, weld cracking or distortion due to residual stresses by coefficient of thermal expansion, segregation weld cracking due to dilution of strengthening aluminium alloys, high solubility for hydrogen when in the molten state leads to weld porosity due to the high heat input, environmental pollution with health hazard associated by emission of fumes and poisonous gases during welding This in turn with the above said problems with fusion welding process look for better technique for welding of high strength aluminium alloys to overcome these problems. while FSW is another form of welding technique by solid state method of joining metals with low heat input without fumes and emissions to become the better choice for welding of these alloys.

1.10 About FSW Process

Efforts to reduce the weight of various vehicles and improve safety performance have resulted in increased application of light-weight aluminium alloys and a recent focus on the weldability of these alloys. Shifting of automotive from steel-intensive body structure to lightweight multi-material structure has demanded new joining processes in the assembly line. The production of complex components using lightweight multi-materials such as advanced high strength steel (AHSS), aluminium, and magnesium alloys requires proper welding methods. In the case of dissimilar welding, steels and non-iron-based alloys differ significantly in terms of chemical, physical, and mechanical properties. Therefore, the conventional joining methods are not normally suitable for joining dissimilar materials. In addition, the use of filler material could alter the properties of the weld joint. Therefore, the next challenge for OEMs was to search for suitable welding methods that could be employed to join dissimilar materials and to obtain joints with adequate strength. In this regard, several solid-state welding methods were introduced that could join dissimilar materials without melting, thus eliminating the problems related to conventional welding methods. Friction stir welding is one of the solid-state joining methods. FSW is a relatively new welding process where a rotating non-consumable tool is used to join two materials, invented by Thomas in 1991 in the Welding Institute (TWI) UK, and proposed to weld lightweight materials, a large variety of dissimilar materials. Since then, FSW has gain popularity in the automotive industry and in many other manufacturing industries due to the flexibility of welding, promising weld quality, no filler wire requirement, no emission fumes and poisonous gases.

FSW uses a non-consumable tool that consists of a shoulder with a large diameter and a pin. Figure 1.3 shows a schematic illustration of the FSW process, depending on the type and thickness of materials. The rotating tool is given a traverse motion along the joining line. As the rotating tool moves along the joining line, the pin profile forces the materials from the front to the rear of the tool, so intermixing them. The softened materials are forged by the mechanical pressure of the tool shoulder to obtain a sound quality weld joint. FSW is free of high heat input and solidification defects. Moreover, in the absence of filler material and fumes produced in the traditional fusion arc welding, the FSW process is not susceptible to defects caused by these factors and is considered to be environmentally friendly (S. B. Aziz et al). The Fig. 1.3 show the Schematic illustration of the FSW process.

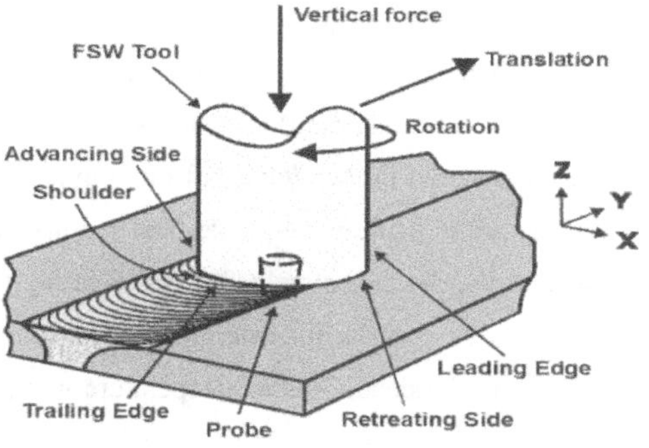

Fig 1.3 Schematic illustration of the FSW process

1.11 FSW Process working steps

The FSW process carried out with the following steps and the Fig. 1.4 (a-d). shows sequence of working process with its stages.

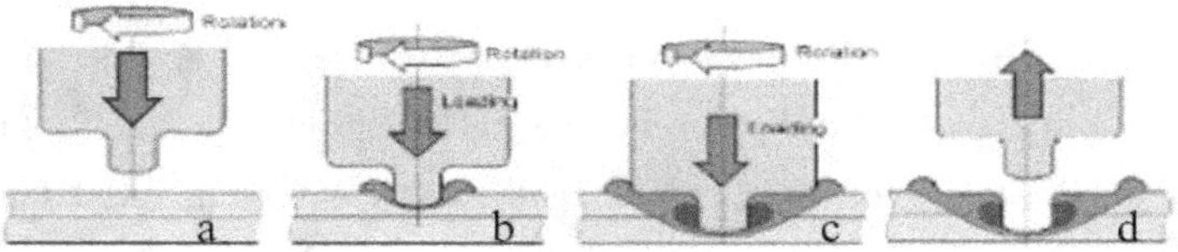

Fig 1.4 FSW working Process with its stages (a) Positioning stage (b) Dwelling stage (c) Welding stage (d) Retracting stage

1.11.1 Positioning and Plunging stage: The tool is brought to the joining line of two (similar/dissimilar) materials. The tool rotating at a constant speed, plunged or inserted through the starting point of the joint line of the workpieces under the action of a vertically downward axial force until the shoulder touches the workpiece surface. This stage initiates the deformation process.

1.11.2 Dwelling stage: The rotating tool, under the action of the axial force and with its shoulder contacting the workpiece surface, is dwelled for 5 to10 seconds depending upon the material and thickness. This dwelling action generates sufficient friction heat at tool workpiece interface and causes plasticization of the workpiece, generating more heat from plastic deformation. The plastic material ahead of the tool ensures a smooth traverse of tool during the subsequent welding stage.

1.11.3 Welding and consolidation stage: The continuous rotating tool, under the action of axial load and with its shoulder enclosing the plasticized volume, is traversed at a constant speed along the desired joint line under the action of a traverse force. The tool rotation continues generating heat by friction, deformation and further important function of material stirring or material flow. The rotation and traverse motions of the tool causes the plastic material in the vicinity of tool pin to flow from the advancing side to the back of the pin to the retreating side. The shoulder applies a forging force on the material behind the pin, effectively filling the cavity formed by the forward motion of pin. This causes material intermixing in the form of atomic diffusion or bonding depending upon the conditions of temperature and pressure, resulting in a proper joint behind the tool.

1.11.4 Retracting and exit stage: Upon arrival at the weld end, the tool is retracted from the workpiece, for cooling, leaving behind an exit hole, several procedures are adopted to avoid the exit hole in the intended weld. These include refilling of the exit hole or using a longer weld length than the intended length etc.

1.12 Friction Stir Welding parameters

The most important FSW parameters are the Tool shoulder profile, Tool pin profile, Tool Tilt angle, Tool Spindle speed, Table traverse or welding speed, Plunge depth. They are reported in the following with the schematic figure 1.5.

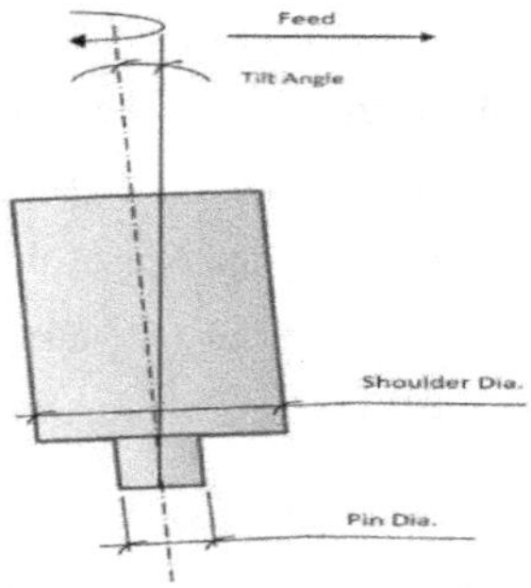

Fig 1.5: Schematic FSW Tool with its parameters

1.12.1 Tool shoulder Profile: Tool shoulder is the large cylindrical part of the tool as shown in Figure 1.6 The tool shoulder prevents plasticized materials coming out of the weldment by applying sufficient pressure.

1.12.2 Tool pin Profile: Tool pin is the small part of the tool, mostly cylindrical in shape, but other configurations of Conical or Taper cylindrical, Threaded cylindrical, Triangular, Square, Pentagonal, Hexagonal and Octagonal, also depends upon the material & thickness of material to be welded. Tool pin intended to shear, removes materials from the front of the tool and displace the contacting surfaces of the workpiece and intermixes at the rear side of the tool. The probe primarily determines the depth of deformation and tool travel speed. which is fully inserted into the welding line during FSW.

1.12.3 Tool Tilt angle: Angle of the tool axis to the normal vector of the workpiece materials such that the rear of the tool is lower than the front, The tool tilt angle normally varies from 0° to 5°.

1.12.4 Tool offset distance: Distance of the centre of the tool pin from the weld line, In dissimilar material welding this parameter plays an important role , depending on the difference in material stiffness and hardness the tool offset comes into picture (E.g. 2.5 mm tool offset at the aluminium side in Copper Aluminium welding).

1.12.5 Plunge depth: The depth up to which the lower surface of the tool shoulder pressed into the workpiece is termed as Plunge depth and it usually varies up to 1 mm. Support is required underneath the tool to react against the vertical down force.

1.12.6 Tool spindle speed

The speed at which the FSW tool rotates with its axis, typically measured in revolutions per minute (RPM).

1.12.7 Table Traverse speed or weld speed

The speed at which the FSW tool moves along the joint line, typically measured in millimetres per minute (mm/min)

1.13 Applications of Friction Stir Welding

The following are the various applications of Friction Stir Welding.

1.13.1 Aerospace: Aluminium is widely used because of its high strength-to density ratio, corrosion resistance, and weight efficiency, especially in compressive designs

- **Aircraft Panels and Structures:** FSW is used in all segments of the aircraft and spacecraft industry to join large panels and structural components in airframes, engines, accessories, and tanks for liquid fuel storage and oxidizers. Towards reducing weight while maintaining strength.

- **Fuel Tanks:** The process is also used to weld fuel tanks for rockets and space vehicles.

1.13.2 Automotive:

- **Lightweight Components:** FSW is applied in the production of lightweight vehicle components, such as aluminium chassis and body panels, to improve fuel efficiency.
- **Battery Enclosures:** The process is increasingly used in the manufacture of battery enclosures for electric vehicles.

1.13.3 Shipbuilding:

- **Hull Sections:** FSW is used to join large sections of aluminium in ship hulls, providing strong and durable joints resistant to corrosion.
- **Deck Structures:** It is also used in the construction of decks and superstructures for ships and boats.

1.13.4 Railway:

- **Train Carriages:** FSW is employed in the fabrication of train carriages, particularly in high-speed trains, where lightweight and strong joints are critical.

1.13.5 Electronics:

- **Heat Sinks and Housings:** The process is used to manufacture heat sinks, housings, and other components in electronics, where precise and reliable joints are required.

1.13.6 Energy:

- **Heat Exchangers:** FSW is used in the fabrication of heat exchangers, which require leak-proof and robust welds.
- **Wind Turbine Blades:** The process is also applied in the production of lightweight and strong wind turbine blades.

1.13.7 Tanks and Armored Vehicles:

- **Hull and Armor Welding**: FSW is used to join aluminium armor plates in modern tanks and armored vehicles. This is particularly important for lightweight military vehicles that require high strength and durability while maintaining mobility.

1.13.8 Missiles and Rocket Motor Casings:

- **Aluminium Alloys in Missile Structures**: FSW is ideal for joining the aluminium alloys commonly used in missile airframes and rocket motor casings. These structures require precision and high strength to withstand the extreme forces and temperatures during launch, flight, and impact.

1.14 Advantages of FSW

FSW is implemented rapidly because it is an energy-efficient solid-state joining technique and shows enormous advantages compared to conventional fusion welding methods. They are,

- A solid-state welding process with no solidification cracking problems.
- An ability to weld thick sections in a single pass.
- The process can be automated, therefore ensuring consistent quality in mass production.
- No edge preparation, shielding gas or filler metal is required.
- Low distortion, shrinkage and residual stresses.
- Excellent mechanical and metallurgical properties.
- Maintaining good dimensional stability and repeatability
- Low energy process. Process efficiency is much higher than many competing processes, particularly laser welding.
- No environmental hazards, improving workplace safety during the process, It does not generates fumes, spatter, or ultraviolet radiation.
- Dissimilar alloy material joints are possible.
- Minimal surface cleaning required
- Very short cooling time, due to the amount of heat generation on weld metal during friction stir welding is comparatively small.

1.15 Disadvantages of FSW

The process has a few drawbacks such as

- Welding speeds can be slower than competing processes, particularly for thin material.
- Workpieces must be rigidly clamped to react against the relatively high forces imposed by the process.
- Initial setup costs may be higher
- Key hole at the end of the joint to be filled separately

- ❖ Size and geometrical limitations of the parts
- ❖ Selection of proper tool with their pin profile according to the job requirements.

1.16 Optimization of welding parameters

Optimization of welding parameters is essential for producing high-quality, reliable, and cost-effective weld joints. It ensures and enhances the quality, reliability, productivity of the outcome weld joint, reduces material and energy consumption, and improves the overall performance and longevity of the welded product. In critical industries like aerospace, automotive, ship building and construction, optimization is key to meeting safety and performance standards while minimizing production costs. Therefore, fine-tuning welding parameters is not just a technical requirement but a strategic advantage in manufacturing. In any welding process, including Friction Stir Welding, proper adjustment of key parameters can significantly influence the final weld's structural integrity, mechanical properties, and overall performance. The correct combination and optimization of these parameters can lead to significant improvements in the following weld qualities:

Mechanical Strength: Optimization ensures sufficient plasticization and mixing of the material, leading to strong weld joints with minimal defects.

Microstructural Refinement: Proper heat control and tool design help in refining the grain structure of the weld, enhancing properties such as toughness and hardness.

Defect Minimization: Issues such as porosity, voids, cracks, and tunnel defects are minimized with the right welding parameters, leading to a more reliable weld.

Fatigue Resistance: Weld joints created with optimized parameters exhibit better fatigue performance due to the uniform distribution of stresses and absence of weak points.

Surface Finish: Good surface finish and minimal flash formation can be achieved by optimizing parameters such as tool rotation speed, traverse speed, and axial force.

1.17 Spiking Neural Networks (SNNs) Optimization Techniques

Spiking Neural Networks (SNNs) are a form of artificial neural networks (ANNs) inspired by the way biological neurons communicate through discrete spikes of electrical activity. Unlike traditional ANNs, which process continuous values, SNNs use spikes (events) to represent information. Due to their event-driven and time-based nature, SNNs can potentially lead to more efficient and biologically plausible computation, especially in energy-limited environments. However, optimizing SNNs poses several challenges due to their discrete spike-based operation and time-dependent dynamics. Welding parameter optimization using Spiking Neural Networks is an emerging area in the field of machine learning and manufacturing processes. Welding is a highly complex process with several controllable parameters such as voltage, current, speed, and feed rate that affect the quality and strength of the weld. Optimizing these parameters is crucial for achieving high-quality welds and minimizing defects.

Spiking Neural Networks differ from traditional neural networks (like deep learning) in that they simulate neurons that "spike" based on discrete events or thresholds, more closely mimicking the way biological neurons operate. SNNs are often more efficient in certain computational contexts, especially when dealing with time-series data or event-driven processes.

1.18 The Main Aims of the Present Research

The main aims of this current research work is to carry out the metallurgical, mechanical, and corrosion properties of AA2219-T87 Aluminium alloy by Friction stir welding using special spherical balls impregnated shoulder enhancement tools of three different pin profiles.

- To conduct the Friction, stir welding of AA2219-T87 Aluminium alloy plates using three different tools of square, hexagonal and octagonal pin profiles with special spherical balls shoulder enhancement to obtain good quality weldments.
- To obtain range of process parameters that can create good quality weldments without defects.
- To carryout macro, micro structural analysis of the FSW joints fabricated out of the previous said three different pin profile tools of special shoulder enhancement towards improvement of mechanical properties (tensile strength, micro hardness)
- To analyze the corrosion behaviour of the FSW joints fabricated out of the previous said three different pin profile tools of special shoulder enhancement for better corrosion properties.
- To select the most efficient favorable pin profile tool amongst three different pin profile tools based on comparative study of their individual performance.
- To implement the optimization techniques in order to obtain the optimal welding process parameters from the various range of parameters there by achieving better

quality weld joints having improved Tensile strength, microhardness and improved corrosion properties.

- To conduct the confirmation test with the optimized process parameters and to compare and confirm the test results
- To conduct a comparative study of best identified spherical ball shoulder enhanced tool pin profile as per the outcome results with the same profile plain shoulder tool and there by establishing the spherical ball shoulder effects on various mechanical, metallurgical and corrosion properties.

1.19 Scope of the Research

- Investigating the newer fabrication techniques, supportive facilities towards development and implementation of advanced materials such as composites, high-strength alloys, and lightweight structures to enhance performance, durability, and fuel efficiency across aerospace, automotive, highspeed railway, shipbuilding structural and other parts.
- Exploring novel manufacturing techniques, to improve production efficiency, reduce costs, performance enhancement and enhance design flexibility.
- Newer FSW tool fabrication technique in evaluating the mechanical strength, microstructural behaviour, corrosion resistance, fatigue resistance, and failure mechanisms of components used in aerospace, automotive, railway, and naval applications to ensure safety, longevity, and reliability.

1.20 Overview of the Thesis

An over view of the five chapters included in the thesis is given below

Chapter-1 -Introduction

This Chapter-1 introduces the current research work on high strength Aluminium alloy material, its various applications and the important production requirements were shared in order to achieve the required properties, the important material joining techniques to fabricate them and their impacts after fabrication in the various properties. This chapter also deals with friction stir welding the one of the ideal joining techniques employed in fabrication of the high strength aluminium alloys, the various essential applications of FSW process, important process parameters, their advantages and disadvantages. It also outlines the scope of the present research.

Chapter-2 - Literature review

This Chapter deals with the overall view of various global level research articles on the joining of high strength aluminium alloy by various welding methods, dissimilar material welding with the high strength aluminium alloys, the mechanical properties, microhardness impacts, corrosion properties, the various other process impacts and their process optimization. This literature review aims to summarize the current understanding of the welding processes and their effects on AA2219 alloy, highlighting advancements, challenges, and key findings to provide a comprehensive overview for future research.

Chapter-3 –Experimental Details

This Chapter-3 narrates the experimental details related to the friction stir welding of AA2219-T87 aluminium alloy with three different pin profile tools. From the raw material to plates the weld specimen preparation as per drawing through , wire cut EDM machine. Friction stir welding tool design and manufacturing of plain as well as special spherical ball impregnated shoulder tools of square pin profile, hexagonal pin profile, octagonal pin profiles are designed. The Friction stir welding on the specimen is made using BISS make friction stir welding machine. The after-weldment test coupons for various inspection were prepared cutting across the joints as per respective ASTM standards by wire cut EDM machine. The macro, micro, EDAX, EBSD analysis of specimen prepared by various polishing process are detailed. The chemical etchant preparation and further etching of various specimens for the respective characterization work are described. The above-mentioned various characterization techniques to analyze the macrostructure and microstructure, microhardness, Tensile testing and corrosion studies were presented in this chapter.

Chapter-4 –Results and Discussion

This Chapter-4 presents the overall details about the results after the friction stir welding of AA2219-T87 Aluminium Alloy joints with three different pin profile tools with special spherical ball impregnated shoulder tools of square pin profile, hexagonal pin profile, octagonal pin profile. First it deals about the visual inspection and X-Ray inspection of friction stir weldments, then about the macrostructure and microstructure details about the various zones of the different pin profile weldments, Then into the micro hardness inspection results of various zones of FSW weldments by the different pin profile tools. Then it details about the Tensile strength results of the various profile tool weldments. Then it shares about corrosion properties of various profile tool weldments. Finally, it deales about Optimization of friction stir welding process parameters using spiking neural network, confirmation test results and comparative studies of various tools performance results and selection of the best tool among the three.

Chapter-5 – Conclusion

This Chapter-5 summarizes the conclusion about friction stir welding of AA2219-T87 Aluminium Alloy joints with three different pin profile tools with special spherical ball impregnated shoulder tools of square pin profile, hexagonal pin profile, and octagonal pin profile tools and the impact on AA2219-T87 weldments. The results reveal the various pin profile tool and process parameters effects on their respective welding joints. After welding joints macro structure, micro structure, grain size and their structures towards the post weld joints namely tensile strength, microhardness, corrosion properties were arrived. The FSW process parameters optimization and their confirmation results are given in detail. The selection of the best tool based on their individual tool profile tool after weldment is results is arrived.

CHAPTER - 2
LITERATURE REVIEW

INTRODUCTION

The AA2219 aluminium alloy is widely used in aerospace, cryogenic applications, and other high-strength structural environments due to its excellent weldability, high specific strength, and superior fracture toughness. Welding processes play a crucial role in determining the mechanical properties and microstructural integrity of AA2219, significantly affecting its performance in critical applications. Researchers have studied various welding techniques, including Electron Beam Welding (EBW), Friction Stir Welding (FSW), and Gas Tungsten Arc Welding (GTAW), to optimize the joint quality, strength, fatigue resistance, and corrosion properties of this alloy. These studies focus on understanding the effects of welding parameters, weld microstructure, and post-weld treatments to improve mechanical behaviour, enhance weld quality, and expand the applicability of AA2219 in demanding engineering applications. This literature review aims to summarize the current understanding of the welding processes and their effects on AA2219 alloy, highlighting advancements, challenges, and key findings to provide a comprehensive overview for future research.

2.1 Effect of Various welding processes on AA2219 alloy

Caiyou Zeng et al. (2024) investigated the microstructure and mechanical properties of 6 mm thick AA2219 aluminium alloy welded using the double-pulsed variable polarity tungsten inert gas (DP-VPTIG) process. The DP-VPTIG process utilizes a variable polarity square wave current, which is modulated simultaneously at ultrasonic frequencies (20–80 kHz) and low frequencies (0.5–10 Hz). The performance of DP-VPTIG welds was compared to those produced by conventional VPTIG welding. The results revealed that the DP-VPTIG welds exhibited alternating bands of fine and slightly coarser equiaxed grains in the weld zone. This hybrid pulsed arc approach significantly refined the grain structure, reducing the average grain size by 51.2% along the transverse section and 61.3% along the longitudinal section compared to conventional VPTIG welds. Furthermore, the morphology of the α-Al + θ-$CuAl_2$ eutectics transformed from a continuous, netlike shape into discrete, granular shapes, improving the segregation of copper solutes. Mechanical testing indicated that the DP-

VPTIG welds outperformed the conventional welds. The average microhardness of the weld zone increased by 8.7% along the transverse section and 5.6% along the longitudinal section. Additionally, ultimate tensile strength, yield strength, and elongation were enhanced by 6.6%, 10.6%, and 20.5%, respectively. These findings suggest that the DP-VPTIG arc welding process offers a promising approach to improving both the efficiency and quality of welded joints through refined microstructures and enhanced mechanical performance.

S. Arunkumar et al. (2018) investigated the mechanical properties of 25 mm thick AA2219 aluminium alloy joined using a double-V butt configuration through the gas tungsten arc welding (GTAW) process. Key welding parameters, including current, voltage, and welding speed, were varied, with tensile strength and microhardness evaluated as performance metrics. A comparison was drawn between the "as-welded" condition and the welds subjected to post-weld aging at 195°C for 5 hours, aimed at improving the mechanical properties. The microstructural analysis provided insights into the variations in mechanical properties. In the as-welded state, the strengthening precipitates had fully dissolved, whereas the aged welds exhibited finer dendritic grains compared to the coarser grains in the as-welded samples. However, the tensile strength of the base metal surpassed both the as-welded and aged welds. The reduction in tensile strength was attributed to the increased size of columnar dendrites and the presence of $CuAl_2$ precipitates in the weld region. Hardness tests revealed that the base metal had higher hardness compared to the as-welded joint, but the weld hardness improved after the aging process. Fractographic analysis of the tensile-tested samples (both as-welded and aged) showed the presence of dimples, confirming a ductile mode of failure.

Sun et al. (2017) investigated the residual stress distributions in stationary shoulder friction stir welds (SSFSWs) and conventional friction stir welds (FSWs) on high-strength aluminium alloy (AA7010). They applied the contour method to map the residual stress over full weld cross sections. Their research demonstrated that SSFSWs, due to a narrower and more uniform heat-affected zone (HAZ), produced slightly lower peak residual stresses compared to conventional FSWs. Additionally, the tensile region in SSFSWs was more focused, attributed to a more concentrated temperature distribution generated by the rotating probe. The study found that increasing the welding speed narrowed the residual stress profile but heightened the peak tensile residual stresses for both processes.

Guo-qing Wang et al. (2017) investigated the impact of weld reinforcement on the tensile behaviour and mechanical properties of TIG-welded 2219-T87 joints. Tungsten inert gas (TIG) welded joints made from 2219-T87 aluminium alloy are commonly used in the construction of fuel tanks for large launch vehicles. Due to the high structural loads these vehicles must endure and weld reinforcement plays a critical role in determining the mechanical performance and reliability of these joints. Both experimental testing and numerical simulations were employed to study the influence of weld toe geometry and positioning on the tensile behaviour of the joints. They demonstrated that weld reinforcement improves stress distribution, thereby reducing the likelihood of stress concentration and enhancing the mechanical properties of the joint. Their findings emphasized that reinforced TIG joints exhibited better tensile strength and elongation than non-reinforced ones with reduced fusion boundary defects. Simulation results demonstrated that variations in the shape of the weld toe can lead to significant differences in elongation with the relative difference reaching as high as 96.9%. Joints with weld toes positioned within the weld metal or in the partially melted zone (PMZ) displayed greater elongation compared to those with weld toes located at the interface between the weld metal and the PMZ. These findings highlight the importance of precise weld toe control to optimize the mechanical properties and ensure the structural integrity of critical aerospace components.

Li et al. (2013) investigated the effects of welding speed on the microstructures and mechanical properties of AA2219-T6 aluminium alloy using reverse dual-rotation friction stir welding (RDR-FSW), a novel variant of friction stir welding (FSW). The study found that defect-free welds were achieved at speeds ranging from 50 to 150 mm/min, with the maximum tensile strength of 328 MPa obtained at 150 mm/min. Microstructural analysis revealed that increasing welding speed led to finer grain structures and a reduction in the softened region, while the minimum microhardness increased with speed. However, at 200 mm/min, cavity defects were observed, resulting in a decrease in mechanical performance. The study emphasized the role of the independently rotating pin and shoulder in controlling heat generation and plastic deformation, which directly affected the weld quality and mechanical properties.

Malarvizhi et al. (2012) analyzed the effects of various welding processes and post-weld aging treatments on the mechanical and metallurgical properties of AA2219 aluminium alloy joints. The study compared three welding processes: gas tungsten arc welding (GTAW), electron beam welding (EBW), and friction stir welding (FSW). FSW joints exhibited superior tensile strength and fatigue properties, with a joint efficiency of 72%, compared to EBW (64%) and GTAW (51%). The post-weld aging treatment at 175°C for 12 hours further improved these properties, with FSW showing the highest tensile strength of 382 MPa after aging. The researchers attributed the improvements in mechanical properties to finer grain structure and the uniform distribution of strengthening precipitates in the FSW joints. Additionally, the fatigue strength of FSW joints was notably higher than that of the EBW and GTAW joints, making FSW the preferred process for fabricating high-strength aluminium structures.

Malarvizhi et al. (2011) conducted a comprehensive study on the fatigue crack growth resistance of gas tungsten arc welding (GTAW), electron beam welding (EBW), and friction stir welding (FSW) for AA2219 aluminium alloy joints. Their investigation revealed that FSW joints exhibited superior fatigue crack growth resistance compared to EBW and GTAW joints. The improved performance of FSW was attributed to its finer, dynamically recrystallized grain structure and more uniform distribution of precipitates. Furthermore, FSW joints demonstrated higher yield strength, tensile strength, and elongation than the other welding methods. These enhanced mechanical properties delayed fatigue crack initiation and slowed crack propagation. In contrast, GTAW joints showed the lowest fatigue resistance due to coarse grain structures and significant reduction in mechanical strength. This study highlighted the critical role of microstructural refinement and mechanical property optimization in improving fatigue performance for aerospace applications.

Malarvizhi et al. (2010) studied the effect of different welding processes on AA2219 aluminium alloy joints. They compared the tensile, fatigue, and corrosion behavior of gas tungsten arc welding (GTAW), electron beam welding (EBW), and friction stir welding (FSW). Their findings revealed that FSW joints showed superior tensile and fatigue properties compared to EBW and GTAW, while EBW joints demonstrated moderate performance. However, the FSW joints exhibited lower corrosion resistance due to the finer grain structure and uniform distribution of strengthening precipitates in the weld zone. The analysis included microstructure evaluations, tensile tests, fatigue testing, and corrosion assessments. Transmission electron microscopy (TEM) revealed that precipitates in the FSW joints were uniformly distributed, contributing to their higher strength, while the EBW joints retained a smaller amount of precipitates. FSW demonstrated the highest joint efficiency, while GTAW exhibited the lowest tensile strength, elongation, and fatigue performance, making it less suitable for critical applications.

Prasad et al. (2010) conducted a detailed study on the fracture toughness (J1C) of electron beam welded AA2219 alloy, focusing on the mechanical properties at the fusion zone (FZ) and heat-affected zone (HAZ). Their research highlighted that electron beam welding (EBW) offered better joint efficiency and higher fracture toughness compared to gas tungsten arc welding (GTAW). The analysis revealed that the FZ and HAZ in T87 temper exhibited higher fracture toughness than the T6 counterpart, with the HAZ being the toughest zone. Additionally, they identified that pulsed current (PC) EB welds showed a slight reduction in toughness compared to constant current (CC) welds. Their findings emphasized the role of tensile testing, Charpy impact testing, and microstructural analysis using scanning electron microscopy (SEM) and transmission electron microscopy (TEM) in understanding the variation in toughness and joint efficiency.

Kanwer S. Arora et al. (2010) explored the effects of process parameters on friction stir welding (FSW) of AA2219-T87 aluminium alloy, comparing it with gas metal arc welding (GMAW). In their study, two identical AA2219 plates (250 × 50 × 5 mm) were welded using an adapted vertical milling machine for FSW and AA2319 filler wire (1.2 mm) for GMAW. GMAW parameters included a travel speed of 6.8 mm/s, 28 V arc voltage, 235 A current, an argon shielding gas, a 16.1 cm/s wire feed rate, and a 12 mm contact tube-to-workpiece distance. The GMAW joint achieved a joint efficiency of 60%, with an ultimate tensile strength (UTS) of 272 MPa and yield strength of 156 MPa. However, the tensile specimen fractured parallel to the fusion boundary, possibly due to partially melted zone (PMZ) cracking, which contributed to reduced strength. For FSW, a Taguchi L9 orthogonal array was used to optimize the process, varying tool rotational speed (N), welding speed (S), shoulder diameter (D), and pin diameter (d). A total of 27 plates were welded using a step pin profile. Axial forces (Fz) were found to be heavily influenced by shoulder diameter and mildly by welding speed and tool rotational speed, while the lateral force (Fx) was mainly affected by welding speed. Tensile strength and percentage elongation of FSW joints were significantly influenced by welding speed and shoulder diameter, with welding speed being the most critical factor. Transmission electron microscopy (TEM) analysis revealed that the lowest

Vickers hardness was found in the weld nugget due to the dissolution and coarsening of second-phase particles. In comparison, the GMAW joint showed liquation in the PMZ, which led to embrittled grain boundaries and a decline in joint strength. The study concluded that FSW offers better mechanical performance and process control than GMAW, with process forces playing a key role in selecting suitable machinery for FSW operations.

Biju S. Nair et al. (2010) conducted experimental studies on AA2219 aluminium alloy plates in two temper conditions: T87 (solution heat-treated, cold-worked, and precipitation-hardened) and T6 (solution heat-treated and precipitation-hardened). The study involved butt welding of these plates (7.0 mm thick, 60 mm wide, and 250 mm long) using electron beam welding (EBW). The variables examined were the temper conditions of the base metal and the mode of EBW employed (pulsed current vs. constant current). The research compared the mechanical properties and fracture toughness of the weld joint, focusing on the fusion zone (FZ) and heat-affected zone (HAZ), against those of the original base material. Findings indicated that EBW resulted in higher joint efficiency and fracture toughness compared to gas tungsten arc welding (GTAW). Among the base metals, the T6 temper exhibited better fracture toughness than the T87 temper. Additionally, the pulsed current (PC) EB welds showed a slight reduction in toughness compared to constant current (CC) welds. The observed reduction in hardness at the HAZ was attributed to the dissolution of strengthening precipitates near the weld interface and over-aging in regions closer to the unaffected base metal. These microstructural changes directly impacted the mechanical performance of the joints.

Srinivasa Rao et al. (1996) investigated the structure-property correlation of auto TIG weldments of AA2219 aluminium alloy used in cryogenic fuel tanks. The study highlighted the effects of post-weld reaging on both the strength and ductility of weldments in the T62 temper condition. Their findings showed that post-weld reaging for 18 hours at 190°C significantly improved the weld efficiency, achieving 80% of the ultimate tensile strength (UTS) and 84% of the yield strength (YS), without a noticeable reduction in ductility. This improvement was attributed to the uniform distribution of $CuAl_2$ precipitates in the heat-affected zone (HAZ). The study emphasized that reaging treatments strengthen the weldments while maintaining mechanical properties, making AA2219 ideal for cryogenic applications. However, welding in the solution-treated condition followed by aging resulted in overaged grain boundary precipitates, leading to reduced ductility in the HAZ.

2.2 Effect of Friction Stir Welding Processes on Aluminium alloy weldments

Bhadra et al. (2024) conducted a comprehensive study on the friction stir welding (FSW) of AA2219 aluminium alloy, focusing on the thermal modelling, microstructural characterization, and mechanical properties. The research revealed that varying welding and rotational speeds significantly influenced the heat generation, peak temperature, and cooling rate, which in turn affected the grain morphology and hardness across different weld zones. Microstructural analysis identified three distinct zones: Weld Nugget Zone (WNZ), Thermo-Mechanical Affected Zone (TMAZ), and Heat Affected Zone (HAZ), each characterized by different grain structures. The Vickers microhardness measurements demonstrated that WNZ exhibited the highest hardness due to dynamic recrystallization and strain hardening. The finite element model (FEM) developed for thermal analysis successfully validated the experimental temperature profiles with an accuracy of 4.86%, showing the effectiveness of FEM in predicting the thermal behaviour of FSW. This study provided valuable insights for optimizing FSW process parameters to improve weld quality.

Lu et al. (2024) investigated temperature distribution during friction stir welding (FSW) of thick 2219 aluminium alloy plates using a least squares support vector machine (LSSVM) model to predict temperature fields. Their study aimed to enhance understanding of temperature gradients in both the thickness and width directions of welded workpieces. The researchers conducted a comprehensive experiment using thermocouples to measure temperatures at various points during FSW, considering welding speeds between 75 and 125 mm/min and rotational speeds from 300 to 450 r/min. The LSSVM model was developed based on these temperature measurements, incorporating factors such as distance from the weld center and the weld's lower surface. The model achieved a maximum relative error of 2.95%, demonstrating high predictive accuracy for temperature distribution. The study concluded that predicting temperature fields using LSSVM could provide valuable data for optimizing FSW parameters, improving weld quality, and minimizing defects caused by temperature-related issues in the welding of thick plates.

Yang et al. (2024) investigated the effects of cold spraying (CS) on reducing residual stress and enhancing the mechanical properties of friction stir welded (FSW) 2219 aluminium alloy joints. Their study demonstrated that applying Cu powder at 400°C and 3 MPa gas pressure resulted in a significant transformation of the maximum residual stress from 159 MPa tensile to -17.4 MPa compressive. The CS process also refined the grain size from 2.1 μm in the FSW joint to 0.7 μm in the cold sprayed joint. Additionally, the proportion of high-angle grain boundaries increased from 38% to 62.7%, and recrystallized grains rose from 19% to 46.9%. These microstructural changes improved the tensile strength of the joint from 305 MPa to 342 MPa, while elongation increased from 3.05% to 6.5%. The study concluded that CS effectively reduces residual stress and enhances joint strength through grain refinement and work hardening, making it a promising technique for improving FSW joints in aerospace applications.

Rathinasuriyan et al. (2024) conducted an investigation into the effect of process parameters on weld geometry and mechanical properties in friction stir welding (FSW) of AA2024 and AA7075 alloys. Their study focused on the influence of rotational speed, traverse feed, and axial force on weld quality. The optimal tensile strength of 351 MPa and hardness of 107 HV were achieved at a rotational speed of 1000 rpm, traverse feed of 30 mm/min, and axial force of 10 kN. Microstructural analysis revealed a dynamically recrystallized stir zone with fine equiaxed grains, enhancing joint strength and hardness. Weld geometry, including bead width and depth of penetration, increased with higher heat input due to increased rotational speed. However, lower feed rates resulted in greater angular distortion, negatively affecting weld quality. This study provided valuable insights for optimizing FSW parameters to improve the mechanical properties of dissimilar aluminium alloys.

Chang et al. (2023) explored the microstructural characteristics and mechanical properties of 2219 aluminium alloy after friction stir welding (FSW) combined with spinning and subsequent T6 heat treatment. Their study revealed that the spinning process significantly refined the grain structure, with the average grain sizes of the base material and the joint measuring 4.7 μm and 6.6 μm, respectively. After T6 treatment, the microstructure was further improved due to the formation of fine needle-like θ' (Al_2Cu) phase precipitates, which effectively hindered dislocation movement, thereby enhancing the mechanical properties. The tensile strength increased to 291.6 MPa in the base material and 331.8 MPa in the welded joint after T6 treatment. These improvements were attributed to dynamic recrystallization and the uniform distribution of strengthening precipitates in the alloy. The study concluded that the combination of FSW, spinning, and heat treatment optimizes both the microstructure and mechanical properties of 2219 aluminium alloy, making it suitable for high-strength applications in aerospace.

Jabraeili et al. (2021) examined the effect of friction stir welding (FSW) parameters on the microstructure and mechanical properties of dissimilar joints between AA2024 aluminium alloy and 304 stainless steel. Their study demonstrated that optimal weld quality was achieved at a rotational speed of 750 rpm, a traverse speed of 65 mm/min, and a zero tool offset. Deviations from these parameters led to defects such as interfacial gaps, tunnel defects, and excessive intermetallic compound (IMC) formation, which negatively impacted joint strength. Microstructural analysis revealed a dynamically recrystallized aluminium matrix and fine ultra-fine-grained steel particles dispersed throughout the stir zone (SZ). The combination of these microstructural features contributed to improved tensile strength and ductility in defect-free joints, while excessive IMC formation led to brittle fracture behaviour.

Dadashpour et al. (2016) investigated the effect of friction stir processing (FSP) parameters on the mechanical properties and fracture behaviour of AZ91C/SiO2 composites. Their study revealed that optimized FSP conditions led to a homogeneous distribution of SiO_2 nanoparticles in the matrix, significantly improving the composite's mechanical properties, such as tensile strength and hardness. Additionally, the microstructure analysis showed a refined grain structure in the stir zone, contributing to enhanced strength. However, the fracture behavior indicated that the inclusion of nanoparticles negatively impacted fracture toughness due to stress concentration points at the particle-matrix interfaces. The study concluded that increasing the number of FSP passes improved the particle distribution, enhancing mechanical properties but compromising toughness due to the formation of voids and micro-cracks around the nanoparticles.

Kadaganchi et al. (2015) investigated the optimization of process parameters for friction stir welding (FSW) of aluminium alloy AA2014-T6 using response surface methodology (RSM). The study explored the impact of tool rotational speed, traverse speed, tilt angle, and tool pin profile on mechanical properties such as yield strength, ultimate tensile strength, and percentage elongation. The findings indicated that the optimal combination of parameters—rotational speed of 1000 rpm, traverse speed of 800 mm/min, tilt angle of 3.5°, and hexagonal tool pin profile—resulted in a maximum ultimate tensile strength of 435 MPa and enhanced elongation. Microstructural analysis revealed refined grain structures in the stir zone due to dynamic recrystallization. The hexagonal tool pin profile was identified as the most effective in promoting material flow and achieving superior mechanical properties, while conical pins yielded lower strength and elongation.

Venkateswarlu et al. (2015) conducted an investigation on the optimization of dissimilar friction stir welding (FSW) of AA2219 and AA7039 alloys. The study focused on determining the effects of tool geometry and process parameters, including rotational speed, welding speed, and axial force, on the mechanical properties of the joints. The researchers found that the optimal combination of parameters resulted in a maximum ultimate tensile strength of 280 MPa and an elongation of 11.5%. Microstructural analysis revealed finer grains on the advancing side (AA2219) compared to the retreating side (AA7039), with inhomogeneous hardness distribution across the stir zone. The study developed a regression model using response surface analysis to predict the effects of process variables and optimize the FSW process for dissimilar aluminium alloys.

Li and Liu (2013) explored the effects of tool rotation speed on the microstructures and mechanical properties of AA2219-T6 aluminium alloy during non-rotational shoulder assisted friction stir welding (NRSA-FSW). Their research demonstrated that defect-free welds could be obtained at tool rotation speeds between 600 and 900 rpm, while cavity defects appeared at 1000 rpm. The optimal tensile strength, reaching 69% of the base material, was achieved at 800 rpm. Microstructural analysis revealed that lower rotation speeds (600-800 rpm) resulted in a basin-type weld nugget zone (WNZ), whereas higher speeds (900-1000 rpm) produced a spherical formation due to excessive heat and plastic flow. The Vickers hardness profile showed softened regions in all joints, particularly in the WNZ and thermo-mechanically affected zone (TMAZ). At 800 rpm, the WNZ exhibited fine, equiaxed grain structures with minimal variation in hardness across the weld. The study concluded that tool rotation speed plays a crucial role in determining weld quality, with 800 rpm offering the best balance between strength and ductility.

Malarvizhi et al. (2011) investigated the effects of three welding processes—gas tungsten arc welding (GTAW), electron beam welding (EBW), and friction stir welding (FSW)—on the mechanical and corrosion properties of AA2219 aluminium alloy joints. Their study revealed that FSW joints exhibited superior tensile and fatigue properties compared to EBW and GTAW joints, primarily due to the finer grain structures and uniform distribution of strengthening precipitates in the weld metal. However, FSW joints showed lower corrosion resistance than EBW and GTAW joints, as the finer grains in FSW led to an increase in intergranular corrosion susceptibility. EBW joints displayed better mechanical performance than GTAW joints but were inferior to FSW joints in both strength and fatigue resistance. This research highlighted the trade-offs between mechanical strength and corrosion resistance in the selection of welding processes for AA2219 aluminium alloy.

Arora et al. (2010) examined the effects of process parameters on friction stir welding (FSW) of AA2219-T87 aluminium alloy. Their research explored how rotational speed, welding speed, and shoulder diameter impact mechanical properties, such as tensile strength and percentage elongation. The study demonstrated that tensile strength decreased with increasing rotational speed, as higher heat input led to dissolution and coarsening of strengthening precipitates. Conversely, increasing welding speed enhanced tensile strength by reducing heat input and thus limiting precipitate dissolution. Furthermore, shoulder diameter significantly influenced the mechanical behavior by affecting heat generation and material flow during welding. Metallographic analysis revealed fine equiaxed grains in the weld nugget, with strain localization and microstructural variations observed across the weld zones. The researchers achieved a maximum joint efficiency of 75%, with their findings indicating that optimal control of process parameters is crucial for improving the weld quality of AA2219 alloy.

Liu Hui-jie et al. (2010) investigated the effect of underwater FSW on the tensile behavior and fracture characteristics of 2219 aluminium alloy. Their study compared joints produced in air (normal FSW) and underwater conditions. The tensile strength of the normal FSW joint reached 324 MPa, which is approximately 75% of the base metal strength, while the underwater FSW joint achieved 341 MPa, about 79% of the base metal strength. The results suggest that reducing the welding temperature through underwater FSW can enhance the tensile strength of the joint. However, elongation decreased to 7.6% for the underwater joint, indicating reduced ductility compared to normal FSW joints. Fracture analysis revealed distinct failure locations for the two types of joints. Normal FSW joints fractured within the heat-affected zone (HAZ) near the transition between the thermo-mechanically affected zone (TMAZ) and HAZ, marking the HAZ as the weakest region. In contrast, the underwater joint fractured at the interface between the weld nugget zone (WNZ) and TMAZ, suggesting an improvement in the HAZ's strength under underwater welding conditions. Hardness measurements further supported these findings, with the normal joint exhibiting a "W-shaped" hardness profile and the lowest hardness recorded in the HAZ (H_V 78–81). Fractographic studies revealed that normal FSW joints had large dimples, indicating significant plastic deformation, whereas underwater joints displayed less pronounced dimples, reflecting reduced plasticity. These results emphasize the potential of underwater FSW to improve joint strength, though at the cost of reduced elongation and plasticity.

C.S. Paglia et al. (2006) conducted a comprehensive study on the microstructure, microchemistry, and environmental cracking susceptibility of FSWed 2219-T87 aluminium alloy. They reported that FSW significantly influences the microstructural evolution of the weld zone, leading to the formation of fine equiaxed grains. The refined grain structure is attributed to the dynamic recrystallization occurring during the welding process, which contributes to enhanced tensile strength and ductility compared to traditional welding methods. Moreover, the study highlighted the importance of the weld zone's microchemistry, noting variations in elemental distribution and the presence of secondary phases that can affect corrosion resistance. The susceptibility to environmental cracking was also evaluated, revealing that FSW joints exhibit improved resistance to stress corrosion cracking (SCC) compared to conventionally welded joints. This improved resistance is partly due to the reduced amount of residual stresses in the FSW joints, which mitigates the driving force for crack propagation.

2.3 Effect of Pin Profile, parameters and other Process on FSW

Yuvaraj et al. (2019) investigated the influence of tool shoulder concave angles and pin profiles on the mechanical properties and microstructural behavior of friction stir welded dissimilar joints between AA7075-T651 and AA6061 alloys. The study employed three concave shoulder angles (1.5°, 3°, and 4.5°) and three different pin profiles (straight cylindrical, square, and triangular). The results indicated that the tool with a 3° concave shoulder and square pin profile produced the highest tensile strength (298 MPa), hardness (143 HV), and bending angle (49°). The square pin's pulsating stirring action facilitated better material flow, leading to fine grains and uniform distribution of strengthening precipitates in the weld nugget zone. The study concluded that tool shoulder geometry and pin profile play critical roles in optimizing the mechanical properties of dissimilar friction stir welded joints by enhancing material flow and controlling heat input.

Mastanaiah et al. (2018) investigated the role of a hybrid tool pin profile in enhancing welding speed and mechanical properties during friction stir welding (FSW) of thick AA2219-T6 aluminium alloy. The study demonstrated that the hybrid tool, consisting of a conical threaded upper section and a triangular cross-section lower section, achieved welds at speeds seven times faster than conventional tools, with a 26% improvement in weld strength. The unique tool design facilitated better material flow and reduced the thermo-mechanically affected zone (TMAZ) softening, leading to improved joint strength and hardness. Microstructural analysis showed finer grains in the nugget zone and a reduction in grain size from top to bottom, which contributed to enhanced mechanical properties. The findings suggested that the hybrid tool allows for higher travel speeds, reduced peak temperatures, and improved material flow, making it ideal for welding thick aluminium alloys.

Bozkurt et al. (2018) explored the effect of tool material properties on friction stir welding (FSW) of AA2124-T4 alloy metal matrix composites (MMC). The study involved the use of three different tools: uncoated, CrN-coated, and AlTiN-coated tools. The results showed that the AlTiN-coated tool produced the highest-quality welds with enhanced tensile strength and wear resistance compared to the other tools. Microstructural analysis revealed fewer defects and a better distribution of SiC particles within the weld zone when the AlTiN-coated tool was used. Additionally, the wear resistance of the tools was directly related to their coating, with the AlTiN-coated tool exhibiting the least wear during the welding process. This study highlights the importance of tool material and coatings in improving weld quality and durability in FSW of MMCs.

Yang et al. (2018) conducted a comparative study on bobbin tool friction stir welding (BT-FSW) and conventional friction stir welding (C-FSW) of Al-Mg-Si alloy (6061-T4) plates, aiming to understand the differences in microstructure and mechanical properties under varying welding conditions. Their research revealed that both processes produced defect-free welds, but BT-FSW had a narrower welding parameter window. Microstructural analysis showed that BT-FSW joints exhibited an hour-glass-shaped nugget zone (NZ), whereas C-FSW joints had a more uniform NZ. Additionally, tensile testing revealed that the maximum joint strength in BT-FSW reached 93% of the base material (BM), comparable to the strength obtained with C-FSW. The hardness profiles in both processes followed a "W" pattern, with the lowest hardness observed in the heat-affected zone (HAZ). The study concluded that while BT-FSW and C-FSW joints showed similar mechanical performance, BT-FSW offered advantages in joining thicker plates due to its balanced axial force and uniform temperature distribution.

Babu et al. (2018) examined the effects of various tool pin profiles on the mechanical and metallurgical properties of friction stir welded joints in cryorolled AA2219 aluminium alloy. Their study employed cylindrical, threaded cylindrical, square, and hexagonal pin profiles. The hexagonal pin profile yielded superior mechanical properties, including higher tensile strength and impact toughness, attributed to the enhanced material flow and dynamic recrystallization induced by the pulsating action of the tool. Microstructural analysis revealed that the weld nugget produced by the hexagonal pin exhibited finer grain structure, leading to improved hardness and fracture resistance. The dissolution of Al2Cu precipitates in the weld zone further contributed to the mechanical strength. Conversely, the cylindrical pin profile resulted in lower joint strength due to insufficient material flow. This research demonstrated that tool geometry significantly influences weld quality, with hexagonal profiles offering the most favourable results.

Padhy et al. (2018) presented a detailed review of friction stir welding (FSW) and its derived processes, focusing on the evolution of microstructures, mechanical properties, and applications. The study highlighted that FSW, a solid-state joining technique, is highly effective for aluminium alloys, offering superior mechanical properties compared to conventional fusion welding methods. The authors discussed various process parameters, such as tool geometry, rotation speed, and traverse speed, and their influence on weld quality. Additionally, friction stir-based daughter technologies like friction stir processing (FSP), friction stir cladding, and friction stir riveting (FSR) were explored, emphasizing their applicability in the automotive and aerospace industries. Padhy et al. emphasized that FSW is gaining prominence due to its energy efficiency, environmental benefits, and ability to join dissimilar materials with reduced distortion and defects.

Kumar et al. (2016) studied the influence of tool materials on the mechanical and microstructural properties of friction stir welded (FSW) 316L austenitic stainless steel butt joints. Two different tungsten-based tools were used: one composed of 99% tungsten with 1% lanthanum oxide (La2O3) and another with 90.12% tungsten, 5.97% nickel, 3.56% iron, 0.22% molybdenum, and 0.15% cobalt. The study found that joints produced with the tungsten lanthanum oxide tool exhibited superior mechanical properties and lower tool degradation compared to those produced with the tungsten heavy alloy tool. The lanthanum oxide tool experienced plastic deformation, while the heavy alloy tool suffered from severe abrasive wear, distributing tungsten particles into the weld zone, which negatively impacted the joint's mechanical performance. The research emphasized the importance of tool material in FSW, particularly for high-melting-point materials like stainless steel.

Rao et al. (2015) studied the influence of tool pin profiles on the microstructure and corrosion behaviour of AA2219 Al-Cu alloy friction stir welds. The investigation involved five tool profiles: conical, triangle, square, pentagon, and hexagon. The results showed that the hexagonal tool profile generated higher peak temperatures during welding, leading to finer grain structures in the nugget zone and increased dissolution of strengthening precipitates like $CuAl_2$. The fine grains and higher dissolution improved the hardness and pitting corrosion resistance of the welds, with the hexagon tool producing the best results in terms of mechanical performance and corrosion resistance. Differential scanning calorimetry (DSC) and electron backscatter diffraction (EBSD) analyses confirmed that the hexagonal profile facilitated a more uniform and refined microstructure, contributing to the superior properties observed in the welds.

Ilangovan et al. (2015) studied the effects of tool pin profiles on the microstructure and tensile properties of friction stir welded dissimilar joints between AA6061 and AA5086 aluminium alloys. Their investigation compared three tool pin profiles: straight cylindrical, threaded cylindrical, and tapered cylindrical. The threaded pin profile produced the best results, yielding a defect-free stir zone, a higher hardness of 83 Hv, and the highest tensile strength of 169 MPa. These mechanical improvements were attributed to the enhanced material flow and fine grain structure achieved by the threaded pin profile. Additionally, the threaded profile reduced the size of the thermo-mechanically affected zone (TMAZ) and heat-affected zone (HAZ), which further improved the joint properties. This study highlighted the significant impact of tool pin geometry on the material mixing, microstructure, and mechanical performance of dissimilar aluminium alloy joints.

Wu et al. (2015) investigated the effects of stationary shoulder friction stir welding (SSFSW) compared to conventional friction stir welding (FSW) in joining high-strength AA7050 aluminium alloys. The study revealed that SSFSW produced narrower welds with reduced heat-affected zone (HAZ) width and more uniform through-thickness mechanical properties. The SSFSW process, with a 30% lower heat input than FSW, exhibited superior cross-weld tensile strength and surface finish. However, SSFSW was found to result in surface speed cracking under certain welding conditions. Thermal modelling supported these findings, showing that the SSFSW process produced a more uniform thermal field and reduced residual stresses. These advantages make SSFSW a promising alternative to FSW for aerospace applications, especially in reducing distortion and improving joint strength.

Kumar et al. (2014) investigated the influence of tool geometries and process variables on the friction stir butt welding of Al–4.5%Cu/TiC in situ metal matrix composites. The study analyzed different shoulder surface geometries and process parameters such as tool rotational speed and welding speed. Their results showed that a flat shoulder surface with concave geometry and a rotational speed of 500 rpm produced welds with a tensile strength reaching 89% of the base material. Microstructural analysis revealed refined grains and redistribution of TiC reinforcements in the stir zone, contributing to enhanced mechanical properties. X-ray diffraction confirmed the presence of intermetallic phases such as $CuAl_2$ and Al_3Ti in the weld zone, which contributed to the observed mechanical performance. This study concluded that the right combination of tool geometry and process variables is essential for optimizing the welding characteristics of metal matrix composites.

Li et al. (2013) investigated the effects of tool rotation speed on the microstructure and mechanical properties of AA2219-T6 aluminium alloy using the external non-rotational shoulder assisted friction stir welding (NRSA-FSW) process. The study revealed that defect-free joints were achieved at rotational speeds between 600 and 900 rpm, while cavity defects appeared at 1000 rpm. Microstructural analysis showed that increasing the rotation speed resulted in grain coarsening, with a basin-type formation in the weld nugget zone (WNZ) at lower speeds and a spherical formation at higher speeds. The maximum tensile strength (69% of the base material) and elongation were observed at 800 rpm. Hardness measurements indicated that the softest regions occurred in the WNZ adjacent to the thermo-mechanically affected zone (TMAZ), and all fractures in defect-free joints occurred at these points. This study emphasized the importance of optimizing rotational speed to achieve superior weld quality.

Huang et al. (2013) investigated the fabrication of a gradient micro-structured surface layer on AA2219 aluminium alloy using in situ rolling friction stir welding (IRFSW). The study demonstrated that this method produced a surface layer approximately 200 μm deep, with microhardness values decreasing from 86.8 H_V at the top surface to 59.4 H_V in the coarse-grained weld nugget zone. The friction coefficient was significantly reduced, and wear resistance was notably enhanced compared to the base material. Furthermore, the corrosion resistance of the micro-structured surface layer was improved, with the corrosion current reduced and the corrosion potential shifted positively. These improvements were attributed to the finer second-phase particles and the gradient microstructure induced by plastic deformation during IRFSW, offering a promising approach for enhancing the surface properties of aluminium alloys used in engineering applications.

Devaraju et al. (2013) investigated the influence of rotational speed and reinforcement particles on the wear and mechanical properties of aluminium hybrid composites fabricated via friction stir processing (FSP). The study optimized the volume percentage of silicon carbide (SiC) and graphite (Gr) reinforcements using the Taguchi method to enhance wear resistance and mechanical properties. Their findings revealed that SiC particles significantly increased microhardness due to their pinning effect, while the Gr particles contributed to reducing wear by acting as a solid lubricant. The hybrid composite exhibited a uniform dispersion of reinforcement particles in the nugget zone, resulting in improved mechanical properties. Additionally, the wear resistance was attributed to the formation of a mechanically mixed layer containing fractured SiC and Gr particles. This study demonstrated the potential of FSP to fabricate surface hybrid composites with enhanced tribological and mechanical properties.

Arora et al. (2012) examined the load-bearing capacity of tool pins during friction stir welding (FSW), focusing on their mechanical behaviour under bending moments and torsion stresses. The study utilized a three-dimensional heat transfer and viscoplastic material flow model to compute the traverse force and torque experienced by the tool pin during FSW of aluminium alloys and steels. The results indicated that shorter tool pins with larger diameters can sustain higher stresses compared to longer, thinner pins. The study also demonstrated that tool pin failure is influenced by both the temperature-dependent shear strength of the material and the welding variables. For materials such as L80 steel, tool pins made from tungsten exhibited deformation and shortening, while pins for aluminium alloys showed better performance with a higher factor of safety. This research provides critical insights into optimizing tool design for FSW of hard alloys.

Malarvizhi et al. (2011) investigated the effects of different welding processes and post-weld aging treatment (PWA) on the fatigue behaviour of AA2219 aluminium alloy joints. They compared gas tungsten arc welding (GTAW), electron beam welding (EBW), and friction stir welding (FSW) processes. Their findings revealed that FSW joints exhibited the highest fatigue strength and superior mechanical properties compared to GTAW and EBW joints, attributed to finer grain structures and better precipitate distribution. Additionally, post-weld aging treatment further enhanced the fatigue strength across all processes, particularly in FSW joints, which showed an improvement in fatigue performance. The study emphasized the role of dynamically recrystallized grains and uniform distribution of fine precipitates in improving mechanical properties and fatigue resistance. These insights provide significant implications for aerospace applications where high fatigue resistance is critical.

Lorrain et al. (2010) conducted an experimental study to understand the material flow path during friction stir welding (FSW) using unthreaded tools. The researchers analyzed the material flow in both longitudinal and transverse sections of the welds, with and without the use of material markers. The results revealed that the material flow with unthreaded tools exhibited similar characteristics to that observed with classical threaded tools, where material was deposited on the advancing side in the upper portion of the weld and on the retreating side in the lower portion. However, the study found a lower vertical material motion towards the bottom of the weld, which was attributed to the absence of threads on the tool. Additionally, the researchers noted that the product of plunge force and rotational speed significantly affected the size of the shoulder-dominated zone, influencing the weld's shape and quality.

Elangovan et al. (2008) examined the effects of tool pin profile and tool rotational speed on the mechanical properties of friction stir welded AA6061 aluminium alloy. The study used five different tool pin profiles, including straight cylindrical, tapered cylindrical, threaded cylindrical, triangular, and square, at rotational speeds ranging from 800 to 1600 RPM. Their findings showed that the joints fabricated using a square pin profile at 1200 RPM exhibited superior tensile strength and microhardness due to the pulsating stirring action and dynamic material flow. Microstructural analysis revealed that the square and triangular pin profiles produced fine, equiaxed grains, enhancing the overall mechanical properties. In contrast, joints made using the cylindrical pin profiles showed inferior properties due to less effective material flow. The results underscored the importance of tool geometry in controlling weld quality and optimizing mechanical performance.

Elangovan et al. (2008) investigated the influence of tool pin profile and welding speed on the formation of the friction stir processing (FSP) zone in AA2219 aluminium alloy. The study evaluated five tool pin profiles—straight cylindrical, tapered cylindrical, threaded cylindrical, triangular, and square—at three different welding speeds. Their findings revealed that the square pin profile at a welding speed of 0.76 mm/s produced defect-free welds with superior tensile properties and higher joint efficiency compared to other pin profiles. Microstructural analysis showed finer grains in the stir zone of joints produced by the square pin profile, contributing to the increased strength and hardness. The study concluded that both tool pin profile and welding speed are critical factors influencing the mechanical properties and quality of friction stir welds in AA2219 aluminium alloy.

2.4 FSW of different dissimilar material joints

Mishra et al. (2024) investigated the rotary friction welding of dissimilar materials, specifically AA2219 aluminium and SS321 stainless steel, incorporating an AA6061 interlayer to mitigate the formation of brittle intermetallic compounds (IMCs). Their study highlighted the inherent radial non-uniformity in microstructural characteristics such as grain size and misorientation, primarily driven by temperature gradients across the interface. The inclusion of an interlayer was found to enhance the joint strength by 31%, primarily by reducing IMC thickness and promoting dynamic recrystallization on the aluminium side. Without an interlayer, pronounced grain refinement occurred due to dynamic recrystallization, while low-angle grain boundaries (LAGBs) on the stainless-steel side indicated dynamic recovery. In joints with the AA6061 interlayer, the interface exhibited less severe deformation and reduced interdiffusion of Fe and Al, limiting the formation of brittle phases. The study concluded that the use of an interlayer, coupled with optimized process parameters, improved both mechanical properties and joint efficiency.

Han et al. (2024) conducted a comparative study on the corrosion properties of 2219 aluminium alloy sheets and additively manufactured (AM) 2319 aluminium alloy. Electrochemical tests and localized corrosion evaluations revealed that the 2219 aluminium alloy sheet exhibited lower corrosion current density and superior corrosion resistance compared to the AM 2319 alloy. The inferior corrosion resistance of the AM 2319 alloy was attributed to the formation of a course, continuous Al_2Cu phase along the grain boundaries. This phase increased intergranular corrosion susceptibility, as confirmed through intergranular and exfoliation corrosion tests. Additionally, the corrosion morphology analysis showed that the AM 2319 alloy developed more severe corrosion pits and intergranular corrosion networks than the 2219 sheet. Heat treatment of the AM 2319 alloy effectively improved its corrosion resistance by dissolving the Al_2Cu phase, reducing intergranular corrosion depth. This study highlighted the need for post-processing treatments to enhance the corrosion resistance of AM aluminium alloys.

Wang et al. (2024) investigated the effects of multi-pass friction stir welding (FSW) repair on dissimilar joints of 2219-T8 and 2195-T8 aluminium alloys, specifically used for fuel tank fabrication. Their study revealed that multi-pass repair FSW significantly impacts the microstructure and mechanical properties of the joints. During the repair process, material flow was altered, enhancing the uniformity in the nugget zone (NZ). Additionally, coarsening of the θ precipitates on the 2219-T8 side resulted in a reduction in hardness at the low hardness zones (LHZs). The tensile strength of the joints decreased with increasing repair passes at room temperature, although no significant reduction in strength was observed at cryogenic temperatures (−196°C). This research highlights the importance of multi-pass FSW for eliminating welding defects, while also pointing out its effects on microstructure evolution and joint performance, particularly in aerospace applications.

Zhao et al. (2024) performed an in-depth study on laser welding of 2219 Al–Cu and 2195 Al–Li alloys, focusing on microstructure evolution and tensile fracture mechanisms. Their research revealed the intricate asymmetrical microstructure formation in the welded joints, with the 2195 Al–Li alloy exhibiting finer grain sizes than the 2219 Al–Cu alloy, attributed to a higher intergranular growth resistance. The distribution of columnar and equiaxed grains, influenced by thermal gradients, was prominent on the 2195 Al–Li side due to the material's lower melting point and higher temperature gradient. Furthermore, the tensile properties of the welds were affected by the distinct element distributions, particularly the depletion of Mg and Li in the 2195 Al–Li side, leading to weaker mechanical properties. This study contributes to the understanding of the role of texture evolution, including the formation of rotated goss textures, in influencing the mechanical performance of the welded joints, offering valuable insights into optimizing welding parameters for aerospace applications.

Li et al. (2024) investigated the tensile fracture mechanisms of heterogeneous friction stir welded joints between 2195 aluminium-lithium alloy and 2219 aluminium-copper alloy. The study revealed that the tensile strength of joints with 2195 alloy on the advancing side (AS) was superior to those with 2219 alloy on the AS, showing 57.3% of the base material's strength on the 2195 side. Using digital image correlation (DIC) and microhardness testing, the researchers found that tensile failure initiated in the shoulder-affected zone (SAZ) of the 2219 side. Cracks propagated through the low hardness zone (LHZ) between the thermo-mechanically affected zone (TMAZ) and the heat-affected zone (HAZ), where eutectic phase accumulation caused stress concentration. The failure mechanism was identified as a mixed ductile-brittle fracture, with brittle cracks forming due to grain boundary concentrations, followed by ductile fractures dominating in the LHZ. These findings emphasize the importance of selecting the optimal advancing side material in friction stir welding to improve joint performance for aerospace applications.

Klimenko et al. (2024) investigated the dissimilar friction-stir welding (FSW) of aluminium alloys 2519, 6061, and 7050 using an additively manufactured tool. The research revealed that the application of an additively manufactured tool, featuring a Tri flute profile and spiral groove, significantly enhanced the mechanical mixing of dissimilar materials, reducing Z-force by nearly 50%. However, despite this improvement, a distinct interface boundary remained within the stir zone, indicating incomplete mixing of the alloys. The study also demonstrated that post-weld T6 heat treatment led to the formation of silicon-rich particles at the interface, which contributed to decohesion and cracking during tensile tests. This finding suggested that post-weld heat treatment might not be suitable for dissimilar aluminum alloy welds due to the development of brittle interfacial phases that compromise mechanical integrity.

Derazkola et al. (2015) investigated the effects of process parameters on the dissimilar friction stir welding (FSW) of AA1100 aluminium alloy and A441 AISI steel. The study evaluated key factors such as tool offset, plunge depth, welding traverse speed, and tool rotational speed. The results indicated that the optimal joint strength, at 90% of the aluminium base metal, was achieved with a 1.3 mm tool offset, 0.2 mm plunge depth, 800 rpm tool rotational speed, and 63 mm/min traverse speed. Microstructural analysis showed that intermetallic compounds formed at the weld interface, contributing to higher microhardness than the A441 steel itself. The study emphasized the importance of controlling tool parameters to avoid excessive heat, which can lead to the formation of brittle intermetallic layers, thus reducing the joint's mechanical properties. These findings demonstrate the potential of FSW in joining dissimilar materials, offering advantages over traditional fusion welding methods in minimizing heat-affected zones and defects.

Chu et al. (2017) examined the effect of local post-weld heat treatment (LPWHT) on the tensile properties of friction stir welded (FSW) 2219-O aluminium alloy. Their study aimed to improve the formability and mechanical properties of large-sized welded structures, where bulk post-weld heat treatment (BPWHT) is impractical. Using resistance heating as a localized moving heat source, LPWHT increased the maximum elongation by 1.4 times compared to as-welded joints and homogenized hardness across the weld zone. Microstructural analysis showed no significant grain size changes in the nugget zone, but grains in the thermo-mechanically affected zone became coarser and more equiaxed after heat treatment. The LPWHT-treated joints displayed tensile strength and elongation values reaching 93.3% and 96.1% of the BPWHT, respectively. Additionally, strain analysis revealed that LPWHT significantly improved strain distribution, especially at 300°C, enhancing the formability of the joints. These findings suggest that LPWHT is a feasible alternative to BPWHT for large-scale welded components.

2.5 Corrosion Behaviour of FSW Aluminium alloy process

Sathiya et al. (2017) conducted experimental research on the mechanical, metallurgical, and corrosion properties of AA2219-T87 alloy using 22.4 mm thick plates. The study compared cryo-treated plates (operating within a temperature range of -90°C to -30°C) with untreated ones. Before undergoing cryo-treatment, the untreated material was subjected to rolling. The cryo-treated samples were processed through cryorolling (CR), achieving a 46% thickness reduction after 37 passes, bringing the final thickness to 12 mm with a true strain rate of approximately $\varepsilon = 3$.The cryorolling process resulted in a 14% improvement in yield strength, hardness, and impact strength, demonstrating that CR is an effective severe plastic deformation (SPD) technique that produces ultrafine grain structures. Corrosion tests revealed that cryo-treated samples exhibited better resistance compared to the untreated material. Specifically, the corrosion current density (Icorr) for the untreated alloy was 0.0925 μA, whereas the CR-treated material showed a reduced value of 0.0332 μA, indicating enhanced corrosion resistance. Furthermore, the impedance value of the cryo-treated material reached 31.8 kΩ cm^2, significantly higher than the untreated material's 13.1 kΩ cm^2. This higher impedance suggests a reduced corrosion rate, highlighting the beneficial impact of cryorolling in enhancing both mechanical properties and corrosion resistance.

Venkata Rao et al. (2015) investigated the corrosion behavior of friction stir welded (FSW) joints fabricated using tools with conical and taper triangular pin profiles. Their analysis revealed that the weld nugget produced by the triangular pin profile exhibited finer grain size and better grain orientation compared to the conical profile. This microstructural refinement improved the material's corrosion resistance, particularly its resistance to pitting corrosion. The enhanced corrosion resistance observed with the triangular pin profile was attributed to improved material flow during the welding process, leading to more effective dissolution of strengthening precipitates in the nugget zone. Additionally, this profile resulted in greater coarsening of particles within the thermo-mechanically affected zone (TMAZ) and heat-affected zone (HAZ), further contributing to better corrosion performance. These findings highlight how tool geometry can influence both the microstructure and corrosion behavior of FSW joints, with triangular profiles offering advantages over conical ones in terms of pitting resistance.

Zhu et al. (2015) conducted variable polarity tungsten inert gas (TIG) welding on AA2219 aluminum alloy to study the impact of post-weld heat treatment (PWHT) on corrosion behavior, metallurgical structure, and mechanical properties. The PWHT involved solution treatment at 535°C for 30 minutes, followed by water quenching, and artificial aging at 175°C for 12 hours. The study revealed that the post-treated weld joint exhibited a more uniform and refined grain structure compared to the as-welded joint. The mechanical performance of the joint also improved significantly after heat treatment, with tensile strength increasing by 44% and joint efficiency reaching 76%. The PWHT weld joint demonstrated superior overall properties compared to the untreated joint. Additionally, corrosion resistance in the heat-affected zones (HAZ) was enhanced due to the separation of Al_2Cu phase precipitates along the grain boundaries. Furthermore, the post-treated weld joint showed better resistance to stress corrosion cracking (SCC) when exposed to a 3.5% NaCl solution. This improvement was attributed to the optimized microstructure achieved through heat treatment, which mitigated the material's susceptibility to corrosion-related failures.

Rambabu et al. (2015) explored the corrosion resistance of friction stir welded (FSW) joints in AA2219 aluminum alloy using various tool pin profiles, including taper cylindrical, taper square, taper triangular, taper pentagon, and taper hexagonal shapes. Their findings indicated that the geometry of the pin tool plays a critical role in influencing both the joint structure and its corrosion behavior. Among the tested profiles, the hexagonal pin tool delivered the highest-quality weld joint, demonstrating superior corrosion resistance compared to the other profiles. To further enhance the joint performance, a mathematical model was employed to predict the corrosion resistance of the welded alloy. Additionally, a simulated annealing algorithm was applied to optimize the FSW parameters and tool profiles, aiming to maximize both tensile strength and corrosion resistance in the welds. These results emphasize the importance of tool design and process optimization in achieving high-performance welded joints for aerospace-grade aluminum alloys.

Xu and Liu (2009) investigated the influence of welding parameters on the microstructure and corrosion behavior of AA2219-O aluminum alloy friction stir welds. Their study revealed that the microstructure and corrosion properties of the weld joint are closely tied to variations in process parameters. A dynamically recrystallized zone with fine equiaxed grains was identified in the weld nugget, indicating significant grain refinement due to the FSW process. The research highlighted that increasing the welding speed from 60 mm/min to 100 mm/min (while keeping tool rotation constant) led to a noticeable increase in the size of Al_2Cu precipitates. Additionally, when the welding speed was held at 80 mm/min, raising the tool rotational speed from 500 rpm to 600 rpm resulted in reduced corrosion resistance. The decline in corrosion resistance was attributed to the formation of coarser Al_2Cu precipitates, which facilitated greater absorption of chloride (Cl^-) ions on the sample surface, promoting corrosion. The weldment exhibited better corrosion resistance compared to the base alloy. This improvement in the joint was likely due to the refined microstructure and optimized precipitation patterns within the nugget, underscoring the significance of balancing welding speed and tool rotation for optimal mechanical and corrosion performance.

C.S. Paglia et al. (2006) conducted a comprehensive study on the microstructure, microchemistry, and environmental cracking susceptibility of FSWed 2219-T87 aluminum alloy. They reported that FSW significantly influences the microstructural evolution of the weld zone, leading to the formation of fine equiaxed grains. The refined grain structure is attributed to the dynamic recrystallization occurring during the welding process, which contributes to enhanced tensile strength and ductility compared to traditional welding methods. Moreover, the study highlighted the importance of the weld zone's microchemistry, noting variations in elemental distribution and the presence of secondary phases that can affect corrosion resistance. The susceptibility to environmental cracking was also evaluated, revealing that FSW joints exhibit improved resistance to stress corrosion cracking (SCC) compared to conventionally welded joints. This improved resistance is partly due to the reduced amount of residual stresses in the FSW joints, which mitigates the driving force for crack propagation.

2.6 Optimization of FSW process

Bhadra et al., (2024), the authors investigated the thermal modeling and microstructural evolution of 2219Al alloys under different FSW conditions. They explored heat-affected zones, such as the weld nugget zone, thermo-mechanically affected zone and the heat-affected zone, through optical and scanning electron microscopy. Microhardness profiles were measured at varying distances from the weld line, and the results were correlated with a 3D finite element (FE) thermal model. This model successfully predicted transient temperature profiles during FSW, showing that increased rotational speed generated higher peak temperatures, leading to significant grain refinement near the weld line. Similarly, the cooling rate and grain size were closely tied to both welding speed and tool rotational speed. Overall, the findings emphasize the importance of parameter control in enhancing weld quality by balancing the microstructural and thermal dynamics.

Logesh et al. (2024) investigated the optimization of friction stir welding (FSW) process parameters for dissimilar AA5052 and AA2219 aluminium alloy joints using Taguchi-based Grey Relational Analysis (GRA). The study aimed to improve the tensile strength of the welded joints by optimizing four critical process parameters: rotational speed, welding speed, axial force, and tilt angle. Using the Taguchi L9 orthogonal array design, nine experimental runs were performed to evaluate the influence of each parameter on tensile strength. The GRA method was used to calculate Grey Relational Grades (GRG), which enabled simultaneous multi-response optimization. The analysis of variance (ANOVA) was conducted to determine the significant factors affecting the GRG, with results indicating that axial force was the most influential parameter, followed by welding speed, rotational speed, and tilt angle. Optimal welding conditions, which included a rotational speed of 1250 rpm, welding speed of 30 mm/min, axial force of 12 kN, and tilt angle of 2°, produced a maximum tensile strength of 153 MPa. The findings of this research demonstrate the effectiveness of using Taguchi-based GRA for optimizing FSW parameters to enhance mechanical properties in dissimilar aluminium alloy joints.

Yamazaki et al. (2022) reviewed spiking neural networks (SNNs) and their various applications in computer vision and robotics. The study highlighted the energy efficiency of SNNs compared to traditional artificial neural networks (ANNs) due to their biologically realistic approach to processing information using spike-based temporal coding. The authors discussed different neuron models, such as Hodgkin-Huxley and Leaky Integrate-and-Fire, as well as synaptic plasticity mechanisms like spike-time-dependent plasticity (STDP). They also provided insights into the challenges of training SNNs, including their non-differentiable nature, and covered the state-of-the-art approaches to implementing SNNs using neuromorphic hardware like True North and Loihi. The review concluded with future directions in the field, emphasizing the need for better training methods and more efficient SNN architectures to improve performance in large-scale applications.

Abualigah et al. (2021) The Aquila Optimizer (AO), initially proposed by Abualigah et al., is a population-based nature-inspired optimization algorithm modelled on the hunting behaviour of the Aquila bird. The AO has gained popularity for solving complex optimization problems across various domains, such as engineering, machine learning, and renewable energy (Buddhadev et al., 2023). Several improvements have been made to the original AO to overcome its limitations, such as slow convergence and premature optimization. For instance, Chaotic Aquila Optimizer, which uses chaotic sequences to enhance search capabilities, has been implemented for optimizing the demand response program in residential microgrids, providing superior performance in comparison to other tested algorithms (Kujur et al., 2023). Similarly, the incorporation of Lévy flight-based strategies to enhance exploration capabilities and mitigate the shortcomings of classical AO in UAV network route planning has proven effective, yielding better results compared to standard metaheuristic methods such as Genetic Algorithm and Ant Colony Optimization (Chaudhari et al., 2023). Additionally, opposition-based learning techniques, such as the combination of AO with Oppositional Based Learning (OBL), have been used for feature selection in machine learning models and enhanced ANFIS predictions, improving the search process and overall efficacy (Al-qaness et al., 2023). Hybridization approaches, including integrating AO with other NIOAs or employing specific algorithmic operators, have been successful in enhancing optimization efficiency (Buddhadev et al., 2023). The rapid development of these AO variations demonstrates its significant adaptability and efficacy in addressing complex and dynamic optimization problems.

Kamal Babu et al. (2018) investigated the friction stir welding (FSW) of cryorolled AA2219 alloy to achieve defect-free welds with optimal strength. Parameter optimization was performed using an artificial neural network (ANN) to model the relationship between the input variables (derived from an L9 orthogonal array) and the resulting mechanical and corrosion properties of the joints. For model training, 80% of the experimental data was used, with the remaining 20% reserved for validating the model's predictive accuracy. The optimal parameters were determined by minimizing the root mean square error (RMSE) between the predicted and actual responses. A genetic algorithm (GA) further refined the FSW parameters, resulting in the following optimal conditions: a tool rotational speed of 1005 rpm, a tool travel speed of 20 mm/min, and a tool tilt angle of 3°. When these parameters were applied in a confirmation experiment, the welded joint exhibited a tensile strength of 333.5 MPa, microhardness of 131 Hv0.5, and corrosion current density of 13.5 μA/cm². Additionally, the dimensions of corrosion pits were minimal, highlighting the effectiveness of the optimized FSW process in producing high-quality joints with enhanced mechanical and corrosion properties.

Babu et al. (2017) conducted optimization of friction stir welding (FSW) parameters for cryorolled AA2219 aluminium alloy to achieve maximum weld strength. The study employed artificial neural network (ANN) modelling in combination with a genetic algorithm (GA) for optimizing parameters such as tool rotational speed, travel speed, and tool tilt angle. The optimal conditions were identified as 1005 rpm, 20 mm/min, and a tilt angle of 3°. ANN was used to model the relationship between process parameters and the mechanical and corrosion properties of the weld joint, while GA provided the optimal parameters to minimize trial costs. The optimized weld joints were subjected to mechanical and corrosion testing, and the results confirmed superior tensile strength and corrosion resistance compared to joints welded using Taguchi L9 orthogonal array parameters. This approach successfully reduced production costs associated with cryorolled plates, highlighting the effectiveness of ANN-GA modelling in welding parameter optimization.

Rambabu et al. (2015) investigated the optimization of friction stir welding (FSW) parameters to improve the corrosion resistance of AA2219 aluminium alloy joints. The study utilized a central composite design with four factors (tool pin profile, rotational speed, welding speed, and axial force) to predict corrosion resistance. The experimental data were analyzed using response surface methodology (RSM), and a mathematical model was developed to predict corrosion resistance. To further optimize the parameters, the simulated annealing algorithm was applied. The findings demonstrated that the corrosion resistance of FSW joints was highly influenced by the tool pin profile, with the hexagonal pin producing the highest corrosion resistance. The study concluded that optimizing FSW parameters could significantly enhance the corrosion resistance of AA2219 alloy, thereby offering benefits for aerospace and defense applications. This research highlights the importance of selecting appropriate welding parameters to achieve desired mechanical and corrosion properties in welded joints.

Teixeira et al. (2012) investigated the application of Multi-Objective Ant Colony Optimization (MOACO) for solving the twin-screw configuration problem (TSCP). TSCP, a combinatorial problem, involves determining the optimal location of available screw elements along a screw shaft, and it is typically approached as a multi-objective optimization problem due to its conflicting objectives. The study compared the performance of MOACO with a previously proposed Multi-Objective Evolutionary Algorithm (MOEA) and a two-phase local search algorithm. The experimental results indicated that MOACO holds substantial potential for addressing TSCP effectively. The authors considered several algorithmic components, including pheromone information, pheromone update strategy, and weight aggregation methods, to enhance MOACO's performance. They found that the best-so-far update strategy for pheromones produced significantly better results than iteration-based updates, and that using multiple pheromone matrices per objective resulted in superior solutions compared to a single pheromone matrix. MOACO outperformed the evolutionary algorithm and local search in approximating Pareto fronts across multiple TSCP instances. Additionally, the MOACO approach for TSCP, which used pheromone relations to define successor screw elements on the screw axis, was superior in terms of finding promising screw configurations, as well as effectively balancing between exploration and exploitation. These findings suggest that MOACO is a suitable algorithm for TSCP, particularly in polymer compounding and other applications involving co-rotating twin-screw extruders.

Vijayan et al. (2010) conducted a study to optimize the friction stir welding (FSW) process parameters for Aluminium Alloy AA 5083 using Taguchi-based Grey Relational Analysis. The FSW process parameters, including tool rotational speed, transverse speed, and axial load, were optimized for multiple responses, namely tensile strength and power consumption. Taguchi's L9 orthogonal array was used to design the experiments, and the study aimed to determine the optimal parameter levels for achieving maximum tensile strength while minimizing power requirements. The grey relational grade was used to identify the optimal parameter levels, and analysis of variance (ANOVA) was employed to determine the significant contribution of each process parameter. The study demonstrated that tool rotational speed significantly affected tensile strength, while axial force was less influential. The optimization revealed that moderate tool rotational speed (650 rpm), low transverse speed (115 mm/min), and lower axial force (9 kN) produced the best results in terms of joint quality. And also highlighted the benefits of the grey relational analysis for handling multi-response problems, converting them into a single optimization problem, which can be effectively validated through confirmation experiments. The confirmation run validated the identified optimal conditions, showing substantial improvement in both the tensile strength and power efficiency of the FSW process. This research demonstrates the utility of combining Taguchi methods with grey relational analysis for multi-objective optimization in welding processes.

Babu et al. (2009) explored the optimization of friction stir welding (FSW) parameters to maximize the tensile strength of AA2219 aluminium alloy joints. Using a central composite design, they analyzed the effects of tool rotational speed, welding speed, axial force, and tool pin profile. The Hooke and Jeeves search technique was employed for optimization, leading to a developed mathematical model predicting tensile strength. Their results showed that the optimal conditions involved a rotational speed of 1627 rpm, welding speed of 0.85 mm/sec, and axial force of 12.25 kN, combined with a square pin profile. This combination produced maximum tensile properties, attributed to efficient heat input, uniform grain structure, and enhanced material flow. The study concluded that optimizing these parameters significantly enhances joint quality, making FSW a reliable method for aluminium alloy applications requiring high tensile strength.

Myhr and Grong (2009) explored the optimization of welding conditions and heat treatment schedules for age-hardening aluminium alloys using a novel modelling approach. The study utilized the general-purpose finite element code WELDSIM, incorporating a natural aging model to calculate the minimum heat-affected zone (HAZ) strength and the equivalent half-width of the reduced strength zone during single-pass butt welding of Al-Mg-Si alloys. Their research demonstrated that strength loss within the HAZ is influenced by factors such as the base metal chemistry, initial temper condition, welding speed, and effective heat transfer coefficient. The authors illustrated that the interplay between these parameters could be condensed into process diagrams, enabling improved design predictions for different combinations of welding conditions. By minimizing the strength loss post-welding, significant weight reductions and cost savings in structural components could be achieved. Myhr and Grong's model also incorporated mechanical predictions of design stress under various loading conditions, which provided a comprehensive simulation of welding-induced transformations and their impact on mechanical properties. The study concluded that optimizing welding conditions and subsequent heat treatment schedules effectively improves the yield strength profile, making WELDSIM a valuable predictive tool for engineering design. This approach offers a more tailored prediction for age-hardening aluminium alloys, improving upon existing empirical methods often used in industry for predicting welding outcomes.

2.7 Overview of Literature

The literature on welding AA2219 aluminium alloy emphasizes the importance of selecting suitable welding techniques and optimizing parameters to enhance material properties. AA2219 is favoured in aerospace due to its excellent weldability, fracture toughness, and strength. Friction Stir Welding (FSW) has emerged as a promising technique, offering superior tensile and fatigue properties compared to Electron Beam Welding (EBW) and Gas Tungsten Arc Welding (GTAW). However, FSW presents challenges such as lower corrosion resistance. Studies highlight the critical role of process parameters, such as rotational speed and welding speed, in determining weld quality. Optimizing these parameters through methods like Response Surface Methodology (RSM) and Taguchi techniques has shown significant improvements in mechanical performance. Tool geometry, specifically pin profile, also influences weld quality, with square and triangular profiles enhancing material flow. Recent advancements in optimization algorithms, including Genetic Algorithms and Artificial Neural Networks, have further improved welding efficiency and joint properties. Overall, optimizing welding processes and parameters is key to achieving high-quality welds, making AA2219 alloy highly suitable for aerospace and other demanding applications.

2.8 Research Gap / Problem Statement

The detailed review of the existing literature on friction stir welding (FSW) of AA2219-T87 Aluminium alloy using various tool pin profiles reveals the following research gaps or problem statements.

- There is a limited body of research exploring the metallurgical and mechanical properties associated with different shoulder geometries with different tool pin profile used for FS welding applications.
- The influence of probe tools featuring distinct shoulder profiles on weld mechanical properties, microstructure, and corrosion behaviour remains largely unexplored
- The effects of spherical ball shoulder design enhancements on FSW weldments particularly in terms of mechanical strength, metallurgical characteristics, and corrosion resistance have not been extensively studied and warrant further investigation.

2.9 Objective of research work

Based on the detailed literature survey on FSW AA2219-T87 Aluminum alloy with various pin profile tools, there is only limited research has been conducted on the FSW different pin profile tools shoulder geometry impacts on metallurgical, mechanical and corrosion properties. Based on this identification of the research gap the following were formulated as research objectives

1. To carryout friction stir welding of AA2219-T87 Aluminum alloy with three different tools of square, hexagonal and octagonal pin profiles with spherical ball shoulder special enhancement tools.

2. To analyze the macrostructure, microstructure, tensile strength, microhardness and corrosion analysis of the after friction stir weldments

3. To utilize optimization techniques to enhance the process parameter selection through Spiking Neural Network and EAO algorithm for the FSW of AA2219-T87.

4. To conduct a series of confirmation tests to verify the effectiveness of the optimized welding parameters.

5. To examine the resulting microstructures and weld zones characterization for three different tool pin profiles of Square, Hexagonal and Octagonal with special spherical ball shoulder

6. Identification of best tool pin profile based on the comparative study of the outcome results of three different pin profile tools. Choosing the best tool for future work.

7. Friction stir welding of plain shoulder best profile tool with the optimize parameter, analysing the micro structure tensile strength

8. To compare the micro structure and mechanical properties of the spherical ball shoulder welded specimens with plain shoulder welded specimens.

2.10 Summary of Literature

The literature on welding AA2219 aluminium alloy emphasizes the importance of selecting suitable welding techniques and optimizing parameters to enhance material properties. AA2219 Aluminium alloy is favoured in aerospace, defence applications due to its excellent weldability, very good cryogenic properties fracture toughness, and strength. Friction Stir Welding (FSW) has emerged as a promising technique, offering superior tensile and fatigue properties compared to Electron Beam Welding (EBW) and Gas Tungsten Arc Welding (GTAW). However, FSW presents challenges such as lower corrosion resistance. Studies highlight the critical role of process parameters, such as tool rotational speed, welding speed and tool tilt angle, in determining weld quality. Optimizing these parameters through methods like Response Surface Methodology (RSM) and Taguchi techniques has shown significant improvements in mechanical performance. Tool geometry, specifically pin profile, also influences weld quality, with triangular, square, hexagonal and octagonal pin profiles enhancing material flow. Recent advancements in optimization algorithms, including Genetic Algorithms and Artificial Neural Networks, have further improved welding efficiency and joint properties. Overall, optimizing welding processes and parameters is key to achieving high-quality welds, making AA2219 alloy highly suitable for aerospace and other demanding applications.

CHAPTER 3

EXPERIMENTAL DETAILS

This chapter deals with the experimental details related to the friction stir welding of AA2219-T87 aluminium alloy and the, raw material , the weld specimen preparation, Friction stir welding, characterization techniques, micro hardness, tensile testing and corrosion studies.

3.1 Materials

The below sections describes about the various aluminium alloys

3.1.1 Aluminium and its alloys

Aluminium and its alloys in the ambient environment have lower density, high strength, high ductility, excellent formability and high corrosion resistance which make them promising candidates for being used extensively in thin food foils, beverage cans, household utensils, cookers to automotive, railway and aircraft structures. In modern life aluminium alloys are the real promising candidates for replacing equivalent steel parts, structures, assemblies and the use of them in the automotive industry is increasing recently, particularly the closure panels such as hoods, decklids and lift-gates. The weldability of aluminium alloys varies depending on the chemical composition of the alloy used.

Wrought sheet of an alloy Al-3Cu-0.8Sn, known as 'partinium' (Minet, 1902, 1905), has been used extensively in the electric vehicle named 'La Jamais Contente' built by Rothschild Coach Builders. Driven by Camille Jenatzy of Belgium, this vehicle attained the first world speed record greater than 100 km/h (105.9 km/h) in 1899 (e.g. Furet, 2008).

In 1906, four years after the publication of Minet's book, Alfred Wilm of Germany discovered the basic procedure by which some aluminium alloys (i.e. the Al-3.5Cu-0.5Mg-0.5Mn alloy that became known as Duralumin) could be heat treated. These techniques which involved precipitation (age) hardening are still used today, for strengthening a wide range of automotive and aerospace components. New advances in technology and understanding that have been developed so as to further the effective utilization of this metal and its alloys. (R.N. Lumley 2011)

3.1.2 Chemical composition of AA2219 alloy

The base AA2219-T87 alloy is chemically composed of Al-Cu-Mn. The received base material is a hot rolled plate and treated with T87 condition. The nominal chemical composition is shown in below Table 3.1. The T87 condition indicates that the material was solutionized at 535˚C, strain hardened by 7%, and then aged for 18 hours between 170˚C and 180 ˚C (Papazian, 1982).

Table 3.1 Chemical composition of AA2219 alloy plate

Cu	Mn	Zr	V	Ti	Mg	Si	Zn	Al
6.30	0.38	0.2	0.10	0.06	0.02	0.20	0.10	balance

Table 3.1 lists the chemical composition of the base metal used in this research work. The base metal is AA2219, an aluminium-copper alloy with high strength and excellent corrosion resistance that is commonly used in aerospace, defence, and marine applications. Each of these elements has a distinct impact on the alloy's qualities. Copper gives the alloy excellent strength and corrosion resistance, while manganese increases the alloy's strength and toughness.

Zirconium is added to the alloy to control the grain size, which improves its strength and hardness. Vanadium and titanium are used to boost the alloy's strength at high temperatures. Magnesium helps to improve the castability and machinability of the alloy, while silicon improves its fluidity during casting. Zinc is added to increase the alloy's strength and hardness, while aluminium is the main component of the alloy and provides it with good corrosion resistance and excellent formability.

Table 3.2 Mechanical Properties of base material AA2219-T87 alloy plate

Yield Strength (MPa)	UTS (MPa)	Elongation (%)	Vickers Hardness @ 0.5 kg load
310	408	23	127

Workpieces in this research were two identical aluminium alloy 2219- T87 plates. Wire cut EDM was used to cut the raw material AA2219 aluminium alloy plates in the sizes of 200X200X 22.7 mm thick into the appropriate dimensions of 100 mm x 100 mm x 6.2 mm.

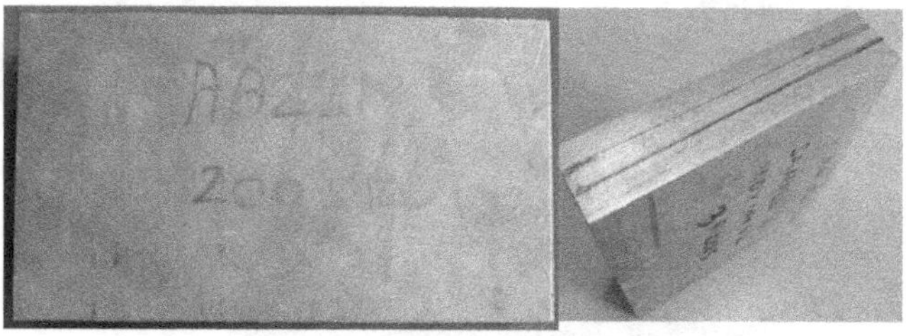

Figure 3.1 Raw material plates of Aluminium alloy AA2219-T87 (200 X200X22.7)

Figure 3.2 EDM Wire cut the weld specimen of AA2219-T87 Aluminium Alloy

3.2 Friction stir welding (FSW)

The figure 3.4 depicts the BISS make FSW machine. In this machine the AA2219 weld specimens were friction stir welded using the three different tool pin profile of Square pin, Hexagonal pin and Octagonal pin profile tools.

Figure 3.3 FSW machine-BISS Make **Figure 3.4** FSW machine with job hold fixture

Figure 3.5 Wire cut weld coupons on FSW fixture

Table 3.3 Biss Friction stir welding machine Specification

BISS -Four Axis Friction stir welding machine Specification		
Axis and Spindle details	**Machine tool parameter**	**Specification Values**
X and Y axis	Load	50 kN
	Stroke Length	500 mm
	Slide Rate of movement	1 to 3000 mm /min
	Accuracy	Plus, or minus 6 microns
Z axis	Load	50 kN
	Stroke Length	300 mm
	Slide Rate of movement	1 to 100 mm /min
	Accuracy	Plus, or minus 6 microns
X and Y axis Servo Motor	Speed	2000 rpm
	Torque	37 Nm
	Power	7.75 kW
X and Y axis Gear Box	Ratio	3
	Backlash	6 Arc min
Spindle servo motor	Spindle Speed	3000 rpm
	Spindle Torque	100 Nm
	Power	30kW
Tool Tilt in Y-Axis	Tilt Angle	0^0 to 5^0
Power Pack	Capacity	4LPM
Control System		UACE2020

3.3 Manufacturing of different FSW tools

AISI H13-DIE STEEL material was chosen of around 40 to 45 HRC to make three separate tools. To obtain accurate weld joint profiles, three alternative pin profiles of a Square, Hexagon, and Octagon were manufactured without draught using a wire cut electric discharge machining equipment and a CNC turning centre. Pin profiles with flat faces (square, Hexagonal and octagonal) are associated with eccentricity. This eccentricity allows incompressible material to pass around the pin profile. Also, three different special shoulder enhancement tools of Square, Hexagonal and Octagonal Pin Profiles with 4mm spherical balls impregnation at mid portion of the shoulder were manufactured for conducting the friction stir welding.

The special shoulder features tool increases the amount of material deformation produced by the shoulder, resulting in increased workpiece mixing and higher-quality friction stir welds. The current research work is to carry out the friction stir welding with special spherical ball shoulder enhancement tools and also with the best plain shoulder tool in order to compare the special shoulder enhancement tool weldments to plain tool weldments. Each tool has specifications with a pin diameter of 6 mm, a pin length of 6 mm, and a shoulder diameter of 18 mm. The below figures shows the actual tool manufacturing for the current FSW AA2219 welding experiment. The various tool profile shapes were depicted in the Figures 3.6, 3.7, 3.8, 3.9 and 3.10

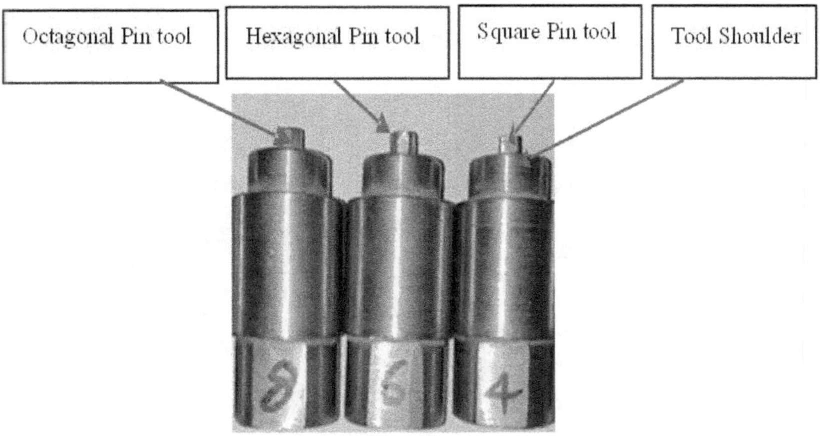

Figure 3.6 Plain Shoulder Octagonal, Hexagonal and Square Pin profile FSW tools

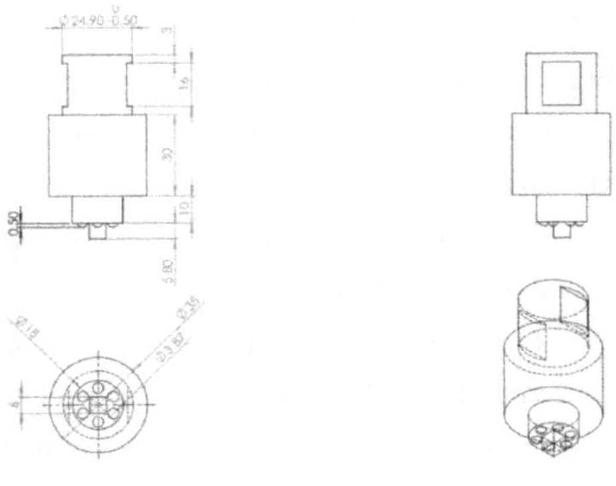

Figure 3.7 Square Pin profile spherical ball Shoulder FSW Tool

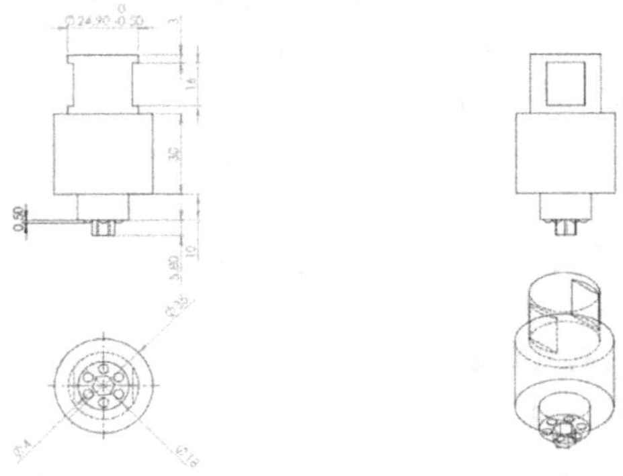

Figure 3.8 Hexagonal Pin profile spherical ball Shoulder FSW Tool

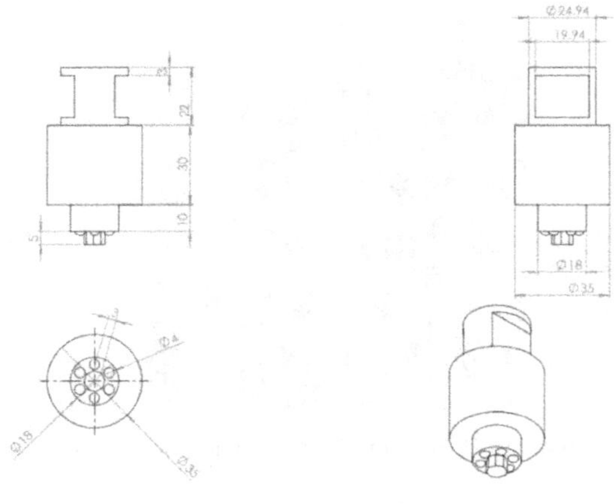

Figure 3.9 Octagonal Pin profile spherical ball Shoulder FSW Tool

Figure 3.10 FSW Tools of (a) Square, (b) Hexagonal and (c) Octagonal Pin profile with spherical ball affixed in shoulder

3.4 Friction Stir Welding of AA2219-T87 Specimens

Workpieces in this research were two identical aluminium alloy 2219-T87 plates. Wire cut EDM was used to cut the raw material plate of size 200X200X 22.6 mm thick AA2219 aluminium alloy plates into the appropriate dimensions of 100 mm x 100 mm x 6.2 mm. FSW joints are constructed using a square butt joint shape, with the plates firmly clamped into the machine table fixture, The fixture was made of high strength steel plates with clamping arrangements to hold the weld. The specimens were fixed in butt joint arrangement. Welding occurred perpendicular to the rolling direction, and a one-pass welding technique was used to construct the joints. The joints were produced using non-consumable tools made of AISI H13 hot die steel as previous mentioned profile tools with following input parameters.

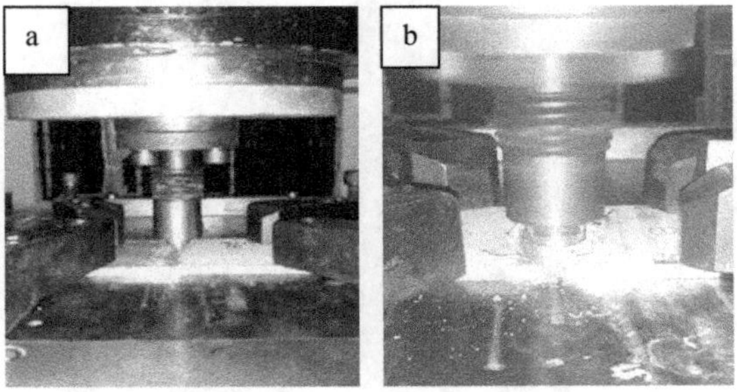

Figure 3.11: (a) AA2219 weld coupons and tool fixed on the FSW machine tool holder ready for FS welding (b) FS welding begins from starting point and going towards another end

The figure 3.11.a shows the AA2219 weld coupons and the tool fixed on the FSW machine tool holder ready for Friction Stir welding and the previous figure 3.11.b shows the Friction Stir welding begins from starting point and going towards another end. The figure 3.12.a shows the hexagonal pin profile, 3.12.b show the square pin profile, and 3.12.c shows the octagonal pin profile weldments

Table 3.4 Taguchi L9 orthogonal array with FSW input parameters

	Input Parameters		
SNo.	Spindle speed in RPM(TSS)	Tool Travel Speed in mm/min (TTS)	Tool Tilt Angle in degrees (TTA)
1	1000	20	2.5
2	1000	25	3
3	1000	30	3.5
4	1250	20	3
5	1250	25	3.5
6	1250	30	2.5
7	1500	20	3.5
8	1500	25	2.5
9	1500	30	3

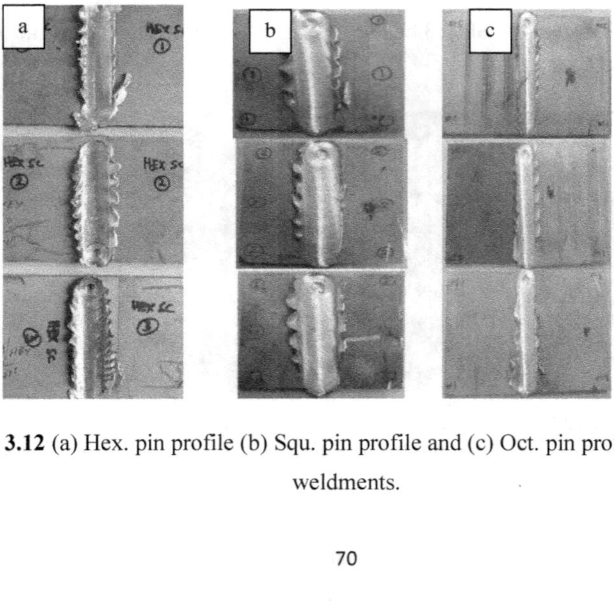

Figure 3.12 (a) Hex. pin profile (b) Squ. pin profile and (c) Oct. pin profile Friction Stir weldments.

3.5 Specimen Preparation for characterization and various inspections

3.5.1 Cutting the Sample

The friction stir welded joint sample is cut from the larger weldment material using a precision cutting method. The cutting process should minimize heat generation and mechanical deformation, as this could affect and alter the microstructure near the surface. Water or coolant is often used during cutting to prevent excessive heat buildup and thermal damage. The below figure 3.13 shows the Wire cut EDM machine setup and the 3.14 shows the Tensile Testing specimen were wire cut ready as per ASTM-E8 -M standards.

Wire cut Electric Discharge Machine (WCEDM) was used to cut the samples across the weld from each tool weld specimens as per the specification for micro hardness inspection, macro structure, micro structural study and Tensile strength analysis.

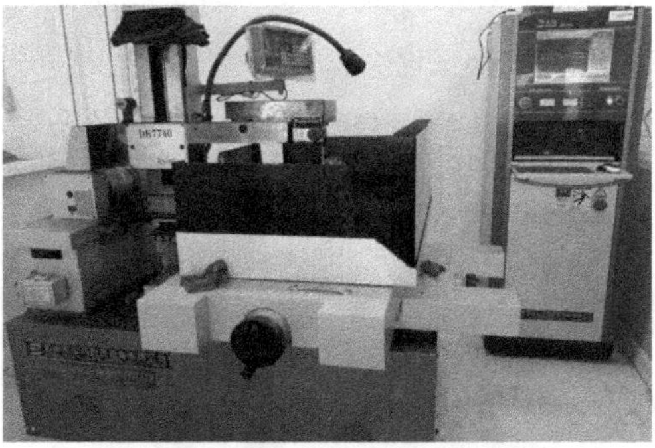

Figure 3.13 Wire cut EDM machine

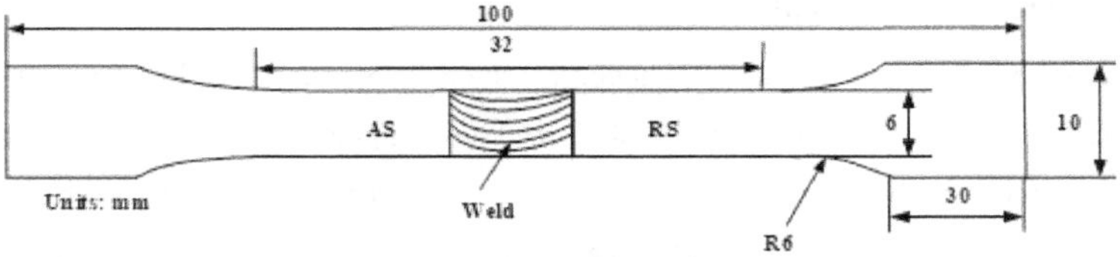

Figure 3.14 Tensile Testing specimen as per ASTM-E8 -M

3.5.2 Grinding machine for Emery polish and cloth polish

An adjustable speed control rotary grinding or polishing machine with rotating turn table were commonly used. These machines have platens, where the emery paper and polishing cloths are mounted. Emery papers are silicon carbide-based grinding papers and available in different grit sizes (180 -1800 Grit size), allowing for precise surface refinement. The below figure 3.13 shows the Emery polish and cloth polish grinding machine.

3.5.3 Coarse Grinding

Small SiC particles that might beglued to the grinding emery paper were sometimes called fixed abrasives. SiC grinding papers were affixed and rotated on a wheel (from 50-3000 rpm). The machine is as shown in below figure 3.18. While in rotation these particles slowly remove the scratch marks from specimen surface.

Start with 120 or 240 grit emery paper was used to remove surface damage or machining marks. Light pressure was applied and the sample was moved in a circular motion. The direction of grinding was changed by 90° to ensure even wear. The sample was pushed face down while polishing, keeping the surface wet by using running water or coolant to reduce heat and flush away debris.

Figure 3.15 Emery polish and cloth polish grinding machine

3.5.4 Intermediate Grinding

The old emery paper was to be changed if the emery surface got clogged with debris, Progress was made to finer grits, such as 400, 600, and 800 grit. These would remove the deeper scratches left by the coarser papers. The direction of grinding by 90° was changed after each grit to check if all the scratches from the previous step were removed.

3.5.5 Fine Grinding

After whole inspection surface became neat and smooth, 1200 grit,1400 grit, or higher were used to achieve a very high smooth, scratch-free surface. The sample was thoroughly washed between each step to prevent contamination from coarser particles. At this point, the surface should be smooth enough for macrostructure inspection under low magnification.

3.5.6 Cloth Polishing

The same rotary machine was typically used, with a soft cloth (usually synthetic, like velvet cloth or microfiber) attached to the rotating platen. An abrasive slurry, such as 6 to 3 -micron diamond paste or 1–3-micron alumina slurry was applied. On the polishing surface water or a lubricant was be sprinkled constantly during the whole process to ensure even distribution of the abrasive and to prevent overheating. The sample was gently pressed against the cloth, moving it in a circular motion. This step removes the fine scratches left on the specimen For fine Polishing a finer cloth with finer abrasive, such as 1-micron diamond paste or colloidal silica was used (around 0.05 microns). Then polishing was continued the surface was scratch-free, with no visible deformation or defects. At this point, the sample should reflect light clearly. This step was critical to achieve the mirror finish required for high-resolution microscopic inspection.

3.5.7 Final cleaning and drying

After the polishing process, thorough cleaning was necessary to remove any residual abrasive particles, or contaminants that could interfere with the etching or microscopic inspection. Rinsing with distilled water was done after polishing. Ethanol or acetone was used to degrease the surface. Then the sample was dried with compressed air or a clean, it was covered with lint-free cloth or fine tissue paper to avoid introducing new contaminants. These steps ensure a scratch-free, highly polished surface ready for macro and microstructure inspection

3.6 Etching for Microstructure Inspection

For microstructure inspection, the polished surface was often chemical etched to reveal grain boundaries, phases, microstructures and other internal features. To characterize the microstructure, standard metallographic procedures were followed as per the ASTM E3-11 standard and to etch the specimens, Keller's reagent was utilized. An optical microscope was employed to characterize the microstructure. For aluminium alloys, Keller's reagent is a common etchant. It is made out of 1 ml hydrofluoric acid (HF), 1.5 ml hydrochloric acid (HCL), 2.5 ml nitric acid (HNO_3) and 95 ml distilled water. The etchant was applied for a few seconds (the etching time was 10−15 s), depending on the alloy and desired structure. Then it was rinsed thoroughly with distilled water and the sample was dried immediately to prevent further corrosion or damage.

3.7 Macrostructure Inspection

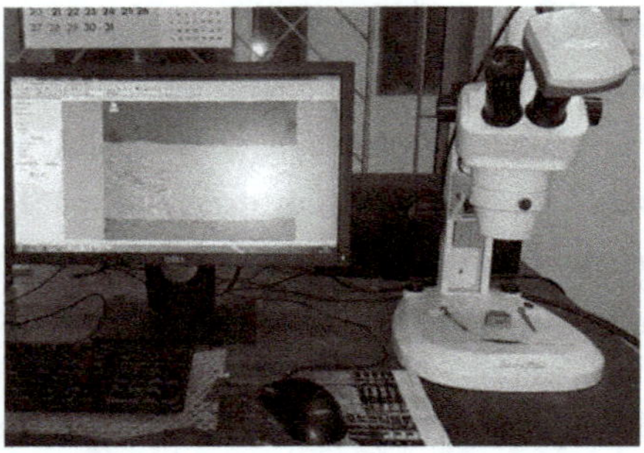

Figure 3.16 Lawrence Mayo Optical Macroscope

Macrostructural inspection refers to the evaluation of the large-scale structural features of a weld, providing insights into overall weld quality, continuity, and the presence of defects. This type of inspection focuses on visible features like weld formation, material flow, and potential flaws that could affect mechanical performance. Optical Microscopy was need for low magnification techniques, (5x–50x) to examine weld zones, Once the surface was prepared through emery polishing and cloth polishing. visual inspection was carried out . To ensure proper good quality the weld was made without any macro-scale defects like voids, cracks, porosity or inclusions and it might reveal layered structures, weld defects, or macro-level grain orientation. The above figure 3.16 shows the Lawrence Mayo Optical Macroscope used in this research for macrostructural inspection.

The bead macrostructure of an AA2219 FSW specimen provides crucial information about the integrity of the weld, the effectiveness of the welding parameters, and the presence of defects. Studying the macrostructure helps in optimizing the process parameters to ensure strong, defect-free joints in critical applications. After etching the sample by examining a cross-sectional macrostructure of the weld bead, the different zones and weld penetration depth on different zones, defect, and the symmetry of the weld were identified.

The metallographic specimen was extracted from the centre of the Friction Stir Weldments. All the joints fabricated in this investigation were analysed at low magnification using Lawrence Mayo optical microscope to reveal the quality of FSW zones. The analysis focuses on to examine various weld zones bead width, detecting surface defects, internal flaws, and ensuring uniformity in the weld structure, which is critical for applications in industries like aerospace and automotive manufacturing.

3.8 Microstructural Zones in FSW of AA2219

The friction stir weld aluminium alloys consists of different zones. The macrostructure of an FSW bead consists of several distinct regions formed by different thermal and mechanical conditions during the welding process. The soften region composed of the weld nugget zone (WNZ), the thermal mechanically affected zone (TMAZ) and the heat affected zone (HAZ) tends to be created by FSW thermal cycles due to the coarsening or dissolution strengthening of precipitates. Fonda RW et. al. (2004), Mishra RS et. al. (2005), Simar A et.al (2008), Starink MJ (2008).

3.8.1 Stir/Weld Nugget Zone (SZ / WNZ)

The material in the central region of the weld bead, was highly plastically deformed and recrystallized due to the stirring action of the tool. The tool rotation and pressure caused significant plastic deformation and dynamic recrystallization. In AA2219, the weld nugget often shows fine equiaxed grains due to this recrystallization process. The microstructure in this zone is often fine and equiaxed. The macrostructure in this region typically shows much smaller, more uniform grains compared to the base material.

3.8.2 Thermo-mechanically Affected Zone (TMAZ)

TMAZ is Located between the stir zone and the heat-affected zone (HAZ). The material here undergoes deformation, but recrystallization may not be complete. Surrounding the weld nugget, this zone experiences both plastic deformation and heating but does not recrystallize completely. The grains in the TMAZ may appear elongated or distorted.

3.8.3 Heat-Affected Zone (HAZ)

HAZ is the outer region near the base material, which experiences heat but minimal plastic deformation. Grain growth or coarsening may occur here. This is the zone where the material has been heated by the friction stir process, but without significant plastic deformation. Some grain growth may be observed here due to the thermal exposure.

3.8.4 Base Material or Un affected Zone (BM/UAZ)

The unaffected area or zone of the parent AA2219 material is far from the welding zone.

The material in this zone is as its of base metal.

3.9 SEM Analysis of FSW AA2219

The Scanning Electron Microscope (SEM) is used for observation of specimen surfaces. In SEM, the specimen is observed at a high magnification in an electron microscope. When the specimen is irradiated with a fine electron beam, secondary electrons are emitted from the specimen surface. Topography of the surface can be observed by two-dimensional scanning of the electron probe over the surface and acquisition of an image from the detected secondary electrons. The specimen should be properly placed and fixed and moved smoothly as required. The specimen stage for a SEM can perform the following movements: horizontal movement (X,Y), vertical movement (Z), specimen tilting (T) and rotation (R). The X and Y movements are used for the selection of a field of view. While the Z movement provides the change of image resolution and the depth of focus.

The electron optical system consists of an electron gun, a condenser lens and an objective lens to produce an electron probe, a scanning coil to scan the electron probe, and other components. The inside of the electron optical system and the space surrounding specimen chamber must be kept at a high vacuum of 10^{-3} to 10^{-4} Pa. Thus, these components are evacuated generally by a diffusion pump. The Figure 3.5 shows the Scanning Electron Microscope set up arrangements.

Figure 3.17 Scanning Electron Microscope with specimens under analysis

When the specimen surface is two-dimensionally scanned by the electron probe, a SEM image appears on the monitor screen of the display unit. Currently, if the scan width of the electron probe is changed, the magnification of the displayed SEM image is also changed. The figure 3.17 shows the Scanning Electron Microscope with specimens under microstructural analysis.

The SEM can provide high-resolution images of the surface topography of the weld, allowing us to inspect the weld bead, microstructure, and any surface defects such as voids, cracks, or porosity in various zones of the weld specimen. The Friction Stir Welded AA2219 specimen provides important insights about bead macrostructure and the quality and characteristics of the weld. FSW bead macrostructure analysis typically focuses on the appearance and distribution of material across the welded region, and it helps evaluate the overall integrity of the weld, detect defects, and assess process and microstructural defects like porosity, inclusions, or microcracks.

3.10 Energy Dispersive X-Ray (EDAX) Analysis of FSW AA2219

EDAX can be used for line scanning or elemental mapping to provide a detailed spatial distribution of elements, across the weld zones. EDAX can reveal critical information about the distribution of alloying elements such as copper, magnesium, and silicon and detect any changes in elemental composition across different weld zones. This helps in visualizing the diffusion or redistribution of elements across the weld cross-section. This helps in understanding the effects of pin profile geometry on material flow, element segregation, and possible defect formation. To analyse, characterize the microstructure and the elemental composition of the AA2219-T87 alloy the Energy-dispersive X-ray spectrometry was used.

3.10.1 Purpose of EDAX in FSW of AA2219 Alloy

- **Elemental Mapping:** EDAX can identify changes in the distribution of key alloying elements such as aluminium (Al), copper (Cu), magnesium (Mg), and other trace elements.
- **Precipitate Analysis:** It helps in understanding the behaviour of precipitates (e.g., Al_2Cu) in different weld zones, particularly the dissolution and re-precipitation processes.
- **Detection of Defects or Contaminants:** EDAX can reveal if any oxidation or contamination occurred during welding, which may affect the performance of the weld.

3.11 EBSD microstructural Analysis of FSW AA2219

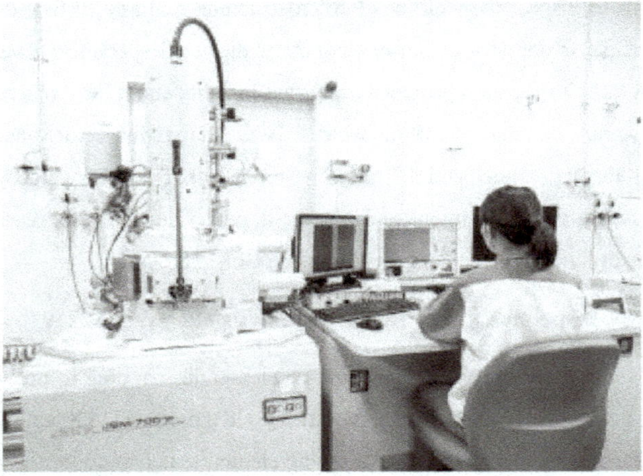

Figure 3.18 EBSD microstructure analyser setup

The figure 3.17 show the EBSD microstructure analyser setup. The EBSD analyser detects the grain geometry that can provide the crystallographic orientation mapping required for the study of recrystallisation. EBSD patterns can also reveal the grain-boundary and sub-grain-boundary features with high contrast at a sufficient level of magnification. The EBSD analysis provides plots for graphical representation of the grain misorientation distribution in the Euler angles system, describing the misorientation probability for the three-dimensional rotation of grains. The EBSD plotting constructs a section of the crystalline texture by irregular restrictive boundaries between the grains. By using the pole figures and their corresponding RGB colour scheme, the crystallographic planes and orientation direction for each individual grain can be locally specified compared to the rest of texture. The colour scheme is based on a colour key triangle named the Orientation Triangle in which each of its corners has a different colour (red, blue, green) assigned to a determined crystallographic orientation, Zhang L (2019), Tamadon A (2019), Fonda R (2007).

3.11.1 Specimen Preparation for EBSD

To obtain accurate EBSD results, the welded specimens need to be carefully prepared, since the quality of the EBSD data depends on the surface finish. Cross-sections of the welded region were cut for microstructural analysis, typically from the center of the weld, which experiences significant plastic deformation and recrystallization.

The cut sections were polished to a mirror finish, often using progressively finer abrasives, followed by polishing with colloidal silica to achieve a surface roughness of around 50 nm or less. The final step may involve electropolishing to remove any deformation induced during mechanical polishing, making the surface suitable for EBSD.

3.11.2 EBSD Scanning and Data Collection

Once the sample was prepared, it was loaded into an SEM (Scanning Electron Microscope) equipped with an EBSD detector. The EBSD scan was performed over the welded region, generating orientation maps that show the crystallographic orientation of grains in the material with details grain structure in different zones of the weld, The orientation of grains helps to determine the texture formed during the welding process, which is linked to mechanical properties like strength and ductility. Misorientation angles: This refers to the angles between grains, indicating the type of boundaries (e.g., low-angle vs high-angle grain boundaries).

3.11.3 EBSD Result Report

After the EBSD data is collected, a detailed report can be generated containing the following: Orientation maps show the crystallographic orientation of grains in different regions of the weld. The Grain size distribution: Average grain size within the nugget zone, TMAZ, and HAZ. The Texture analysis: Analysis of the preferred grain orientations in the weld region, which affects anisotropic properties of the material. Misorientation angles and boundaries: Information about the grain boundary types and their distribution, which affects the mechanical properties like tensile strength and ductility. The specimens were prepared by mechanical polishing followed by vibration polishing. For the above analysis by EBSD, only the specimens used for the metallographic structure observation were etched. The influence of the input welding parameters (TSS, TTS, TTA, Tool Profile) on the microstructure can be assessed by examining the grain refinement, grain morphology, and texture development across different weld regions were analysed.

3.12 Microhardness Inspection of FSW AA2219

Vickers micro hardness measurements were carried out with Wilson Hardness testing machine model-Vickers -402 MVD as per ASTM E384 with a load of 0.5 kg with a dwell period of 10 seconds.

Vickers micro hardness values were taken across the joint interface at the centre of the welds to identify the difference in micro hardness formation of various zones. The unwelded base parent metal in its initial T87 condition showed a hardness of HV 125. The indents were spaced at distance of 0.3 mm in accordance with ASTM - E384. The microhardness inspection revealed the distribution hardness values along the weld centreline through cross-section. The below figure 3.19 shows the Vickers Hardness Testing machine with sample and diamond indentation image on various zones with the respective zones microstructure background.

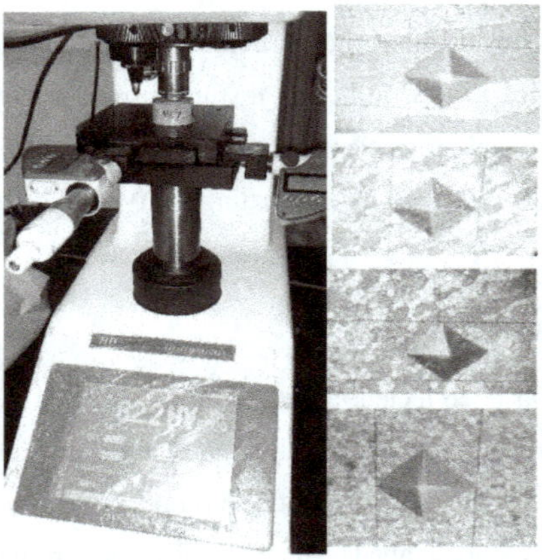

Figure 3.19 Vickers Hardness Testing machine and diamond indentation

3.13 Corrosion Analysis of FSW AA2219

Potentiodynamic polarization tests were applied to determine the electrochemical corrosion behaviour of weldment samples. The Ivium Potentiostat/Galvanostat is commonly used in electrochemical research for conducting a variety of tests, including potentiodynamic polarization studies. Potentiodynamic polarization is a fundamental electrochemical technique to study the corrosion behavior of metals and alloys. Using an Ivium potentiostat for potentiodynamic polarization tests involves controlling the potential of the working electrode (the sample being tested) and measuring the resulting current to determine corrosion parameters. The below figure 3.20 shows IVIUM electrochemical workstation setup.

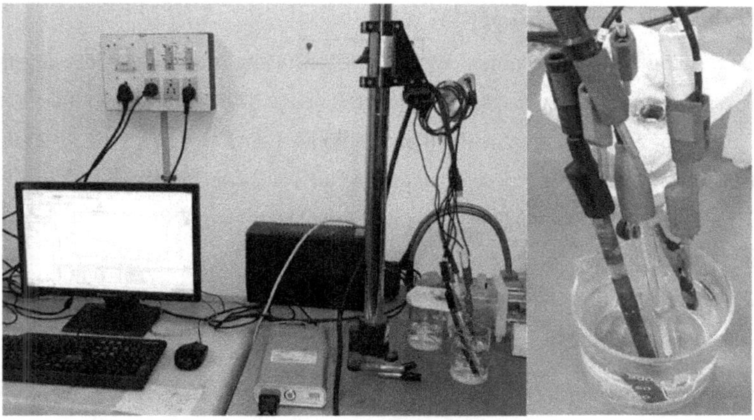

Figure 3.20 IVIUM electrochemical workstation setup

The IVIUM electrochemical workstation was used to carry out the corrosion experiments. Tests were carried out in a three-electrode electrochemical cell containing the electrolyte with the FS Weld sample, a platinum rod, and saturated calomel as the working electrode (anode), counter electrode (cathode), and reference electrode, respectively. Initially, the base alloy was immersed in the electrolyte solution for 30 min and the open-circuit potential was established. The range of the potential was fixed as –1.9 to 0.5 V and the scan was conducted at a speed of 0.1667 mV s–1. To test each specimen, a fresh electrolyte solution mix was used. FS Weld surface was prepared according to the ASTM E3-11 standard. and cleaned using acetone. An electrolytic solution comprising 3.5 wt% sodium chloride (NaCl) in demineralized water was prepared with the aid of a magnetic stirrer. A testing area of 36.00 mm^2 was exposed to the electrolyte.

AA2219, is often employed in aerospace applications due to their high strength, excellent thermal stability and good performance at cryogenic temperatures. The primary concern for such alloys, especially in welded form, is corrosion, which can affect the long-term durability and structural integrity of the components. AA2219 is an Al-Cu alloy that typically contains around 6.3% Cu, with small additions of other elements such as Mn, Zr, and Ti. The high copper content makes it stronger but also more prone to corrosion, particularly in marine or aerospace environments where moisture and chloride ions are prevalent.

3.14 Tensile strength Analysis of FSW AA2219

Tensile strength of the friction stir weldment specimens were measured using the Tinius Olesen Tensile Testing machine model- (H50KL). The figure shows the same machine set up for Tensile testing of Friction Stir Welded AA2219 alloy using a Tinius Olsen tensile testing machine is an important method to evaluate the mechanical properties of the material, especially its strength and ductility. Tensile testing of FSW AA2219 is done to determine the performance of the alloy in different regions of the weld and base material, which may have distinct mechanical behaviours due to the friction stir welding process. The following were evaluated by the tensile test.

- **Ultimate Tensile Strength (UTS)**: The maximum stress that the material can withstand while being stretched.
- **Yield Strength (YS)**: The stress at which the material begins to deform plastically.

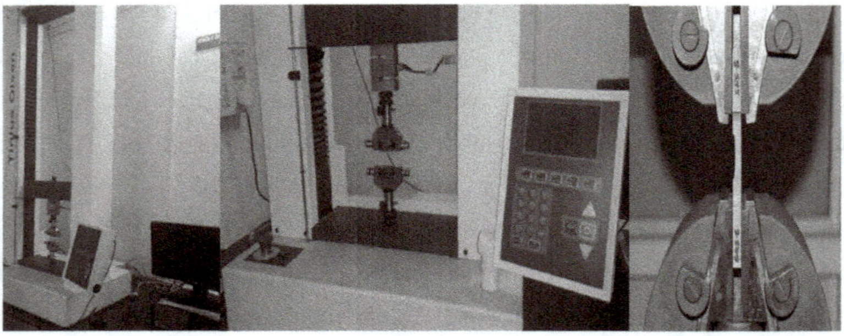

Figure 3.21 TINIUS OLSEN Tensile Testing machine setup with specimen under grip

- **Elongation at Break**: The amount of strain the material can withstand before breaking.
- **Fracture Location**: Identifying whether failure occurs in the weld zone, heat-affected zone, or base metal.

The tensile samples were prepared asper the ASTM-E8 Standard. The job grips were used to hold the tensile specimen properly. The crosshead moves at a set speed rate to apply tension to the weld specimen, an extensometer is used to measures elongation of the sample with high accuracy. The machine having 50 kN load cell, measures the force applied to the sample. The above figure 3.21 show TINIUS OLSEN Tensile Testing machine setup with specimen under grip for tensile examinations. The below figure 3.22 and 3.23 show the across weldment wire cut hexagonal profile tool and square profile tool tensile specimen ready for Tensile examinations.

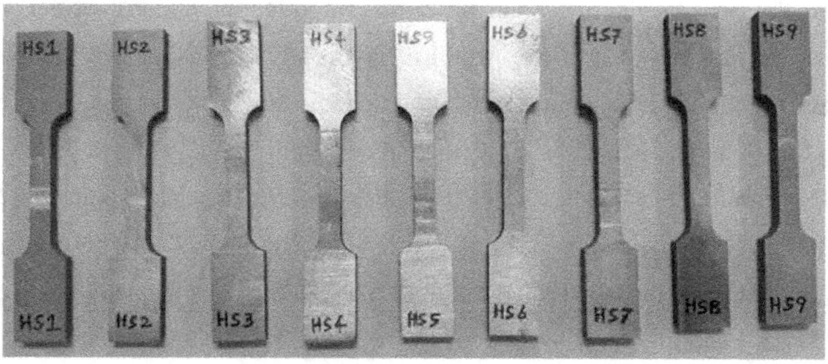

Fig.3.22 Across weldment wire cut hexagonal profile tool tensile specimen ready for Tensile examinations

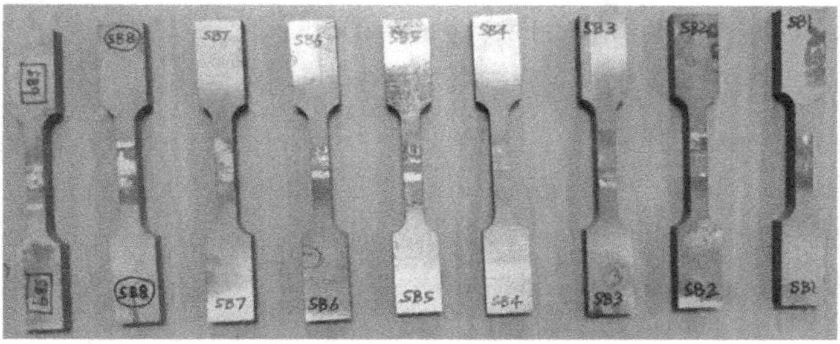

Fig.3.23 Across weldment wire cut square profile tool tensile specimen ready for Tensile examinations

3.15 FSW Process Parameter Optimization

Optimization of welding parameters is essential for producing high-quality, reliable, and cost-effective weld joints. In order to ensure the quality, reliability, productivity of the outcome weld joint, selection, setting and usage correct welding parameters is a must. The process parameters must be carefully optimized for each specific application, considering factors like material type, joint configuration, and desired weld properties. The control of these parameters in Friction Stir Welding is crucial to achieving high-quality, defect-free welds. By optimizing rotational speed, traverse speed, axial force, tool tilt, and other factors, produce strong, durable joints in a variety of materials, making FSW a versatile and valuable welding technique.

3.16 Overview of Spiking Neural Networks (SNN)

Spiking Neural Networks are a biologically inspired class of artificial neural networks that mimic the way the human brain processes information. Spiking neural networks represent a special class of artificial neural networks (ANN), inspired by the biological neural networks found in the brain. Unlike traditional neural networks that use continuous activation functions, the behaviour of biological neurons is modelled by a particular kind of artificial neural network called SNN. It process information using spikes, which are discrete events. SNN examines specific brain structures and assesses changes in transmission and axonal states throughout time. The SNN delays the activation of certain neurons, which raises the quality value of the neurons. Neighbouring neurons' threshold values can rise or fall when a neuron is activated, depending on the situation. The SNN structure uses coder and decoder algorithms to calculate these values numerically, where neuron models communicate by sequences of spikes. Networks composed of spiking neurons are able to process substantial amount of data using a relatively small number of spikes. The SNN works on the following means:

1. Data Collection: Gather data on welding parameters such as Tool rotational spindle speed, welding speed –Table Traverse speed, Tool Tilt Angle, downward pressure/load, and Tool Probe Profile, Pin diameter, Shoulder diameter, Shoulder Profile etc. This dataset will serve as the training data for the SNN.

2. Preprocessing: Preprocess the data to normalize it and remove any noise or outliers that could affect the performance of the SNN.

3. Model Architecture: Design the architecture of the SNN. This involves determining the number of layers, the number of neurons in each layer, and the connectivity between neurons. Since SNNs are event-driven and work based on the timing of spikes, careful consideration should be given to the temporal aspects of the data.

4. Training: Train the SNN using the pre-processed data. Use various training algorithms, such as Spike Prop or Super Spike, which are specifically designed for training SNNs.

5. Optimization: Once the SNN is trained, it can be used to optimize welding parameters. The SNN takes input data related to the welding process and generates output spikes that represent the optimized parameters.

6. Validation and Testing: Validate the performance of the optimized parameters using real-world welding experiments. This step helps ensure that the optimized parameters generated by the SNN result in improved welding quality and efficiency.

3.16.1 SNN Parameter Optimization

In the context of FSW of AA2219 alloy, SNN can be used to simulate the link between the welding specifications and the strength of joints that are welded. In the model, each neuron takes into account each signal that is sent to it but is not instantly activated. In order to process the neuron's signal, it must successfully catch the threshold that has been established for triggering. The coding system used by the SNN supports this increase or decrease signals. Here, the signal's pulse frequency and pulse range must both be considered by the model. Next comes the decoder event, where each code corresponds to a series of integers. The LIF algorithm's mathematical formula is provided in Equation (1).

$$R(M) - \frac{U_M(M)}{S_M} = P_M \frac{\partial UM(M)}{\partial M} \qquad (1)$$

The neurons' activity is the SNN's output. The activity represents the firing or non-firing of the neurons over time. In the context of FSW of AA2219 alloy, to forecast the welded connections mechanical characteristics based on the welding settings. the neurons' long-term firing or inactivity. The least root means square error (RMSE) value calculated from Equation. 2 in the context of the FSW of the AA2219 alloy is selected as the best as a consequence of the SNN modelling utilizing the three methodologies.

$$MSE = \sqrt{\frac{1}{N}\sum_{j=1}^{N}(A - F)^2} \qquad (2)$$

Where, A is Actual value, F is forecasted value N is the number of experiments were included in the data set. Several studies found that the tanh function produced better results than the other function, hence used as activation in both the hidden and output layers. Equation. 3 shows the hyperbolic tanh equation.

$$U(y) = \frac{1 - \exp(-zy)}{1 + \exp(-zy)} \qquad (3)$$

A deep network architecture is the SNN architecture. In SNN, the input layer, hidden layers, and output layer make up the model structure as shown in the below Fig 3.24

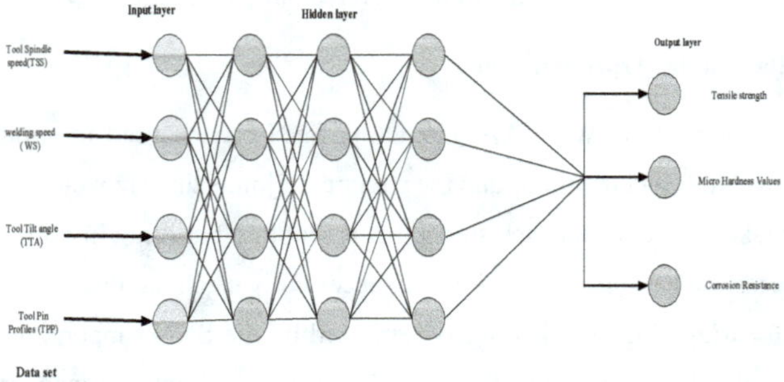

Figure 3.24 SNN architecture with FSW input and output process parameters

3.16.2 Equilibrium optimizer (EO)

The motivation mathematical model and the method for the EO are presented in this section. The EO method was built on the foundation of the straightforward well-mixed dynamic mass balance on a control volume, which employs a mass balance equation to describe the concentrations of a nonreactive ingredient in a control volume as a function of its various attributes. The mass balance equation provides the underlying physics for the conservation of mass entering, leaving, and forming in a control volume. An expression for the first-order ordinary differential equation's general mass-balance equation.

A process is known as "mass changing in time" when the amount of mass entering the system plus the amount being generated inside the system less the amount leaving the system equals the amount of mass changing in time.

$$U\frac{T_O}{T_u} = FO_{bs} - FO + M \qquad (4)$$

Within the control volume, O stands for concentration, U stands for control volume F is the rate of volumetric flow into and out of the regulating volume, and $U\frac{T_O}{T_u}$ is the total mass change rate contained within the controls region. M is the rate of mass generation inside the control container, and O_{bs} stands for the concentration in a situation where there is no generation.

When reaching zero, $U\frac{T_O}{T_u}$, a stable equilibrium condition is attained. Rearranged in some way (1) permits the reciprocal of being a resident time, represented here as or the renewal rate λ, $\frac{F}{r}$ to be used to solve for $\frac{T_O}{T_u}$ as a function of $\frac{F}{r}$. (i.e., $\lambda = \frac{F}{r}$). In the control volume, the concentration (O) in relation to time (e) may then be solved using a different arrangement of Equation (5) as follows:

$$\frac{T_O}{\lambda\, O_{bs} - \lambda O + \frac{M}{U}} = T_u \tag{5}$$

Eq. (6) illustrates how Eq. (7) has been $iter$ grated through time:

$$\int_{O_t}^{O} \frac{T_O}{\lambda\, O_{bs} - \lambda O + \frac{M}{U}} = \int_{e_t}^{e} T_u \tag{6}$$

$$O = O_{bs} + (O_t - O_{bs})X + \frac{g}{\lambda u}(1 - X) \tag{7}$$

X is determined as follows in Eq. (8):

$$X = Exp[-\lambda(e - e_t] \tag{8}$$

where e_t and o_t are depending on the incorporation period, the beginning of the time, and the level of focus. Equation (7) can be used to estimate the concentration in the controlled volume using a known rate of turnover, or it can be used to compute the total rate of turnover using a simple linear regression with a known production rate and other parameters.

In this part, EO is built from the aforementioned equations as a generic framework. In EO, a concentration and a particle's location in the PSO algorithm are comparable to a solution and a particle, respectively. The updating criteria for a particle are presented in three terms, Equation (7) shows that each particle updates its level using three distinct terms. The optimal concentration is a solution drawn at random from a pool of best-so-far solutions. A concentration differential between a single particle and its equilibrium level is connected with the additional term, which acts as a direct search method.

3.16.3 Function evaluation and initialization

EO, like a great deal of state of play-heuristic algorithms, begins its optimization process with the starting population. After uniform random set up in the search space, the first concentrations are generated depending on the number of elements and measurements:

$$O_x^{initial} = O_{min} + Rand_x(O_{max} - O_{min}) \qquad x = 1,2, \ldots .n \qquad (9)$$

$Rand_x$ is a range of random vectors [0, 1], $O_x^{initial}$ is the ith particle's initial concentration vector, O_{min} and O_{max} stand for values for the dimensions are the smallest and largest, and the number of particles in the population as a whole is n. They are sorted initially, and then their capacity for function is assessed to choose the candidates for balancing.

3.16.4 Candidates and the pool of candidates in equilibrium (O_{bs})

The method finally converges on a state of equilibrium, which is meant to reflect the optimal overall state. The state is unknown at the start of the optimum process, thus only stability candidates are chosen to provide the elements with a search pattern. These contenders are the top four performers among the entire refining process' detected particles, plus an additional particle, based on several works under various types of case issues. These four substitutes help EO boost its exploration ability, while the average helps with harvest. The value is arbitrary and determined by the parameters of the optimization problem. Other numbers of candidates may be used. To update the locations of the other wolves in the optimization algorithm, various strategies can be employed.

One common approach is to perform local search around the best candidate solutions found so far. When there are fewer than four candidates, the technique performs worse in bidirectional and composition functions but better in unimodal functions. If there are more than four applicants, the result is going to be overturned. To create the pool of equilibrium states, these five particles, which are suggested as candidates, are used:

$$\vec{O}_{bs,pool} = \{\vec{O}_{bs(1),}\vec{O}_{bs(2),}\vec{O}_{bs(3),}\vec{O}_{bs(4),}\vec{O}_{bs(ave),}\} \qquad (10)$$

During an iteration, the concentration of each particle varies by selecting at random applicants from among those chosen with the same probability. For instance, the first particle may update all of its concentrations based on $\vec{O}_{bs(1)}$ in the initial iteration, and then according to $\vec{O}_{bs(ave)}$ in the second. each particle is gone through an updating process, with each particle receiving roughly the same number of updates for each feasible solution.

3.16.5 Exponential term (X)

The increasing term follows, supplementing the primary concentration updating rule (X). The precise definition of this phrase is to assist EO to achieve an appropriate equilibrium of development and exploration. Because a real control volume's turnover rate may fluctuate over time, is taken to be a random vector in the $Range$ $[0, 1]$.

$$\vec{X} = e^{-\vec{\lambda}(e-e_t)} \qquad (11)$$

where time, e, is an established iterative function $(Iter)$ and, as a result, becomes shorter the more repetitions there are:

$$e = (1 - \frac{Iter}{max - iter})^{\left(A_2 \frac{Iter}{\max\_iter}\right)} \qquad (12)$$

where $Iter$ and Max $iter$ are the current and maximum iterations, respectively, and A_2 is a constant value used to regulate the capacity to exploit. Verify convergence before proceeding. In addition, this work investigates how to speed up exploration and exploitation while slowing down search time.

$$\vec{e}_t = \frac{1}{\vec{\lambda}} In\ (-A_1 sign\ (\vec{v} - 0.5) \left[1 - E^{-\vec{\lambda}e}\right] + e \qquad (13)$$

where the constant A_1 governs the capacity for exploration. The stronger the capacity for exploration, the lower the exploitation performance, and the higher the A_1. Experimentation ability increases while exploitation ability diminishes, as A2 increases. The third element, sign $(\vec{v} - 0.5)$, affects how exploration and exploitation are conducted. R represents a random vector having a 0 to 1 value range. A_1 and A_2 are equal to 2 and 1, respectively, for all of the issues that are later addressed in this work. Eq. (14) shows the revised version of Eq. (11), with the substitution of Eq. (13) into Eq. (11)

$$\vec{X} = A_1\ Sign\ (\vec{v} - 0.5) \left[E^{-\vec{\lambda}e} - 1\right] \qquad (14)$$

3.16.6 Rate of generation (g)

One of the most essential components of the suggested approach for delivering the precise answer by increasing the exploitation phase is the production rate. Multiple models can be employed in applications of engineering to portray the rate of production as a function of time. One general model, for example, defines the rate first order exponential decline method of manufacturing, as seen below:

$$\vec{g} = \vec{g}_t E^{-\vec{Y}(e-e_t)} \tag{15}$$

where g_t denotes the starting point and Y denotes the decay constant. This work makes the assumption that $Y = \lambda$ and use the previously identified exponentially term in order to have a more controlled and systematic search pattern and to reduce the number of random variables. The following is the final set of growth rate equations:

$$\vec{g} = \vec{g}_t E^{-\vec{Y}(e-e_t)} = \vec{g}_t \vec{X} \tag{16}$$

$$\vec{g_t} = \overrightarrow{gom}(\overrightarrow{O_{aQ}} - \vec{\lambda}\vec{O} \tag{17}$$

$$\overrightarrow{gom} = \begin{cases} 0.5v_1 & v_2 \geq gm \\ 0 & v_2 < gm \end{cases} \tag{18}$$

where the random integers $v_1$and v_2 are in the $range$ $[0, 1]$ and the gom The array of values is constructed by using the same value that is received from Equation (18). gom, The calculation takes into account the updating process' possible benefit from the creation term.

Generation Probability, a different number, is what determines the likelihood of this contribution, Eq. (14) and (15) specify the Gom process by which of this contribution.

For instance, if gom is 0, g is equal to zero and all of that particle's dimensions are updated without the use of a generation rate term. At *gm* = 0.5, The equilibrium between both exploration and extraction can be achieved. Finally, here is the EO updating rule:

$$\vec{O} = \vec{O}_{aQ} + (\vec{O} - \vec{O}_{aQ}).\vec{X} + \frac{\vec{X}}{\vec{\lambda}U}(1 - \vec{X}) \tag{19}$$

where U is considered as a unit and X is provided in Equation (11). In order to select the right location, the second term is in responsibility of conducting a thorough investigation of the area. This term makes a greater contribution to exploration and so makes use of significant concentration fluctuations. By identifying a point, the third phrase aids in increasing the correctness of the response. Hence, the generation rate term (Eq.19) which controls concentration benefits from little variations in concentration and contributes more to exploitation. a variety of variables such as particles levels, balance applicants, and the rate of turnover (λ), The signs of both the second and third terms may be identical or opposed.

The third term's variance is increased by low rates of turnover (e.g., 0.05) in the denominator, which also aids in the investigation of particular aspects. The second term in Eq is represented by $O_1 - O_{aQ}$ (19). whereas $\lambda O_1 - O_{aQ}$ denotes the third phrase (g is the function of g_t). The generation rate terms Equations govern how these variances change. Since every dimension is different, this enormous variance mainly affects the parameters with low values.

3.16.7 Hybrid equilibrium optimization with Aquila optimizer (EAO)

The proposed EAO algorithm introduced by the aquila strategy to equilibrium optimization to improve the optimization performance to prevent the equilibrium fall on local optima or terminate the condition before executing overall iteration therefore, the parameters of the equilibrium optimization are taken into consideration to provide better results using aquila strategy. The following shows the mathematical derivation of the aquila optimizer Equation. (20)

$$C_3(i+1) = \left(C_{Best}(i) - C_m(i)\right) \times A - ran + \left((p_o - l_o) \times Rand + l_o\right) \times \delta \quad (20)$$

where $C_3(i+1)$ is the result of the third search procedure for the solution of i in the following iteration (C_3). Until ith iteration, the probable position of the prey is shown by $C_{Best}(i)$.

The mean value of the current solution at the ith iteration is indicated by $C_m(i)$. Between 0 and 1, the value "rand" is random. A and δ are the adjustment parameters. p_o is the upper bound and l_o is lower bound for the problem the thermal error data value in this work is set to a negligibly tiny value of 0.1 because it is too modest. Because the value of the thermal error data in this work is too low, it has been adjusted to a negligibly small value of 0.1.

The above mentioned aquila based equation is derived from the empanelled exposition strategy. Based on this strategy the hybrid strategy emerged by introducing the parameter of the equilibrium EAO algorithm. The mathematical formulation of the EAO algorithm is shown below.

$$\vec{O} = \left(\vec{O}_{aQ} - C_m(i) \times A - rand + (O_{max-}O_{min}) \times rand + O_{min}\right) \times \delta \quad (21)$$

The EAO algorithm can be used to improve the friction stir welding process's parameters and boost the strength and quality of the connection. The following relationship is use to determine the intermediate coded values.

$$V_y = 2\,[2V - (V_{Max} + V_{Min})]/(V_{Max} - V_{Min}) \quad (22)$$

If V_y is the variable's necessary coded value, V represents any value of the variable between V_{min} and V_{max}, V_{min} being the variable's lower limit and V_{max} being its upper limit.

3.16.8 Proposed methodology

To optimize the Welding Parameters of AA2219 alloy is by using SNN and EAO algorithm. SNN is well-suited for controlling systems that involve the processing of time varying signals. The EAO algorithm can be used to search for the optimal process parameters and output characteristics such as tool rotation spindle speed in rpm, Tool Traverse speed or Welding Speed in mm/min, Tool Tilt Angle in degrees, in order to maximize the UTS, Microhardness, Corrosion resistance of the welded joint.

The algorithm would start with an initial set of parameter values and iteratively update them based on the objective function of RMSE. Equilibrium Aquila optimization (EAO), a new optimization technique, can be utilized to get past the issues with the FSW process and improve the performance of the AA2219 alloy. In order to better handle the problems in FSW applications, the EA optimization approach combines the ideas of equilibrium optimization (Mahajan et al. 2021) and Aquila optimization (Abualigah et al. 2021).

EA optimization can be used to improve the tensile strength and elongation characteristics of welded joints by reducing the creation of intermetallic compounds, voids, and porosity during the welding process. Equilibrium Aquila optimization used in achieving the desired mechanical properties by providing optimum input process parameters.

3.16.9 Developing FSW Window

Developing a welding window for FSW of AA2219 alloy involves identifying the optimal range of process parameters that produce consistent, high-quality welds while minimizing defects and other issues.

The welding window for FSW of AA2219 typically involves a range of process parameters such as Tool spindle speed-TSS, Welding speed-WS or Tool traverse speed TTS), TTA-Tool tilt Angle and tool geometry. The parameters are selected based on the properties of the material and the desired weld quality. The following list contains the basic processes that make up in the equilibrium optimization algorithm.

1. Randomly establishing the Population
2. Objective Function Assessment
3. Fitness function finding.
4. Up until the method converges to the ideal point, apply genetic operations such as crossover, mutation, and random selection iteratively. It used the goal Function to carry out the SNN optimization process.
5.Dimensions of the population produced at random: 100.
6.Crossover percentage: 0.9. mutation rate: 0.01.
7.Random selection is a selection criterion.

The following relationship is used to determine the intermediate coded values. The rolled AA2219 plates of size 100 mm by 100 mm by 6.2 mm thick were used as the basic raw material, To determine the upper and lower bounds of the process parameters for the 6.2 mm thick aluminium alloy AA2219, trial runs were carried out. The parameters' practical limitations were selected such that the joint should be devoid of obvious flaws. The greatest and bottom limits of the parameter were also coded as +2 and -2, respectively. The below table 3.5 shows the various welding parameters and their levels, going to be used during the friction stir welding and in parameter optimization process.

Table 3.5 The levels of the weld Parameters used in the FS welding process

Symbol	**Weld Parameters**	**Units**	**Levels**		
			1	**2**	**3**
X	Tool Spindle Speed (TSS)	rpm	1000	1250	1500
Y	Table Traverse Speed (WS)	mm/min	20	25	30
Z	Tool tilt Angle (TTA)	degree	2.5	30	3.5

3.17 Summary

This chapter details the experimental setup for friction stir welding (FSW) of AA2219-T87 aluminium alloy, focusing on material preparation, welding, and testing. The alloy's chemical composition, known for strength and corrosion resistance, was cut into dimensions using wire-cut EDM. FSW was performed using a BISS machine with square, hexagonal, and octagonal tool pin profiles. Specimens were analyzed for microhardness, tensile strength, and microstructural features, using techniques like grinding, polishing and etching. Vickers microhardness tests, tensile testing, and corrosion analysis via potentiodynamic polarization were conducted. Advanced tools such as SEM, optical microscopy, and EDAX were employed to assess weld quality and elemental distribution. Optimization of the process was achieved through Spiking Neural Networks (SNNs and Equilibrium Aquila Optimizer (EAO), identifying ideal welding parameters to enhance mechanical properties and corrosion resistance, while minimizing defects.

3.18 Workflow charts

The comprehensive overview of the research, detailing the materials utilized and the methodologies employed for the experimental procedures and characterization, is depicted in the Figure 3.25.

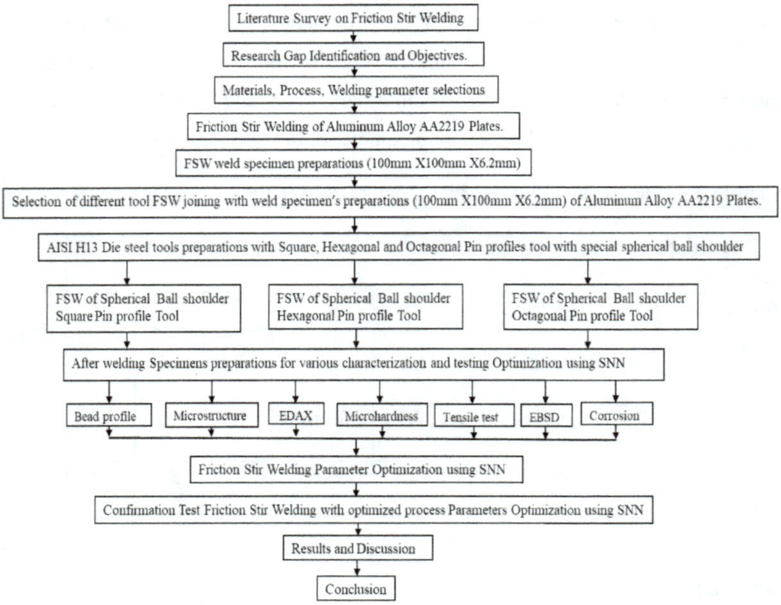

Fig 3.25 Research workflow chart

3.19 Objective based research Methodology

- AA2219-T87 Aluminium alloy plates using different tools of square, hexagonal and octagonal pin profiles with special spherical balls shoulder impact towards enhancement of better UTS , micro hardness and Corrosion strength.
- Refining of micro structure in TMAZ by the introduction of shoulder balls impregnation in different probe tools which alters the metal constituent towards better outcome.
- Impact of spherical ball enhancement on the shoulder of the tool towards alteration of grain orientations of the FSW weldments tool shoulder underneath portions for future special scope applications.

The comprehensive experimental research workflow, illustrated in Figure 3.25, provides a structured overview of the study's key objectives. This figure highlights the raw materials utilized, complete with their specifications, along with the specially enhanced FSW tool profiles designed for optimized friction stir welding. Additionally, it outlines the methodologies employed for result analysis and characterization, facilitating the selection of the most effective FSW tool. The ultimate goal is to contribute to future industrial applications by refining material fabrication processes for critical components.

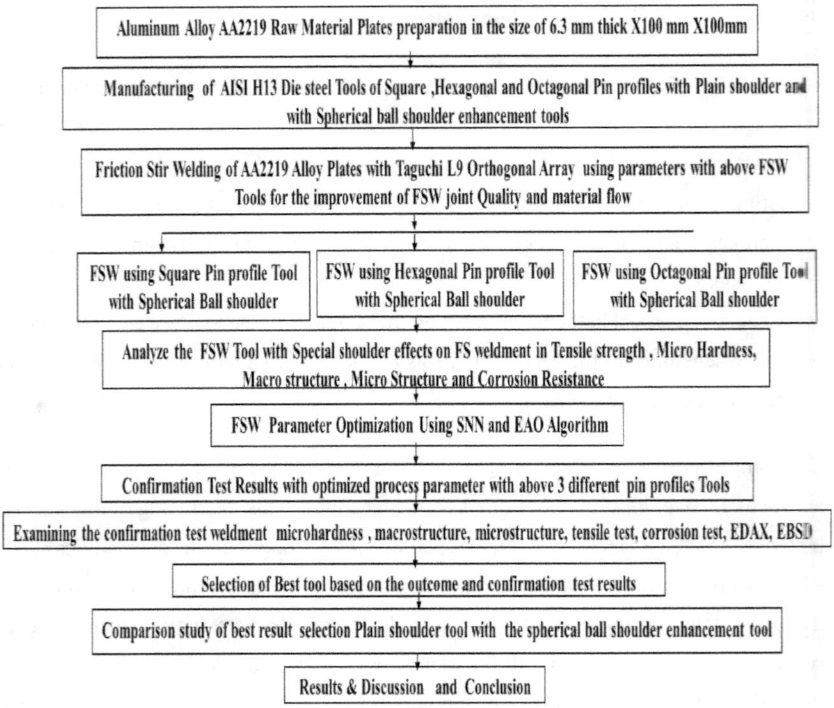

Fig 3.26 Experimental objective research methodology flow chart

CHAPTER 4

RESULTS AND DISCUSSIONS

After the successful friction stir welding of AA2219-T87 Aluminium Alloy joints with three different pin profile tools with special spherical ball impregnated shoulder tools of square pin profile, hexagonal pin profile, octagonal pin profile in BISS friction stir welding machine. The after-weld joints specimens were wire cut as per the ASTM Standards for the Tensile testing, Micro hardness inspection, macro and microstructure analysis and further chemical characterization.

The chapter is divided into five major portions to present detailed result analysis. First it shares about the visual inspection, Then into the micro hardness inspection results of various zones of FSW weldments by the different pin profile tools. Then about the Tensile strength results of the various profile tool weldments, then macrostructure and microstructure details about the various zones of the different pin profile weldments, then it shares about the corrosion properties of various profile tool weldments. Finally, it shares about optimization of friction stir welding process parameter using spiking neural network, confirmation test results, comparison studies of various tools performance results and selection of the best tool among the three were shared.

4.1 Visual inspection

For initial welding, three samples in each tool were welded in order to understand the quality of the joints. Visual inspection was carried out for if any poor quality due to lack of bonding, pin holes, kissing bonds, voids, cracks etc. Before conducting further all samples friction stir welding, various all characterization work, visual inspection of weld joint was carried out to ensure that all welded samples were defects free. The joints should not have lack of bonding, if any micro cracks, voids, kissing bonds, too much thickness reduction, over flash generation etc. On visual inspection of the initial FS weldments were visually found OK.

4.2 Microhardness Inspection of friction stir weldments of square, hexagonal and octagonal pin profile tools of AA2219-T87

The unwelded base parent metal in its initial T87 condition was measured at several locations with an average micro hardness value of 125H_V. The weld metal's micro hardness value on the various zones after friction stir welding were significantly decreased. In comparison to various weld zones, the exterior parent metal that was not got affected by the welding thermal cycle's, the hardness of the AA2219-T87 FSW zone was lower. The central WNZ parts of the weldment had the lowest hardness of all the weldment areas. The highest hardness value was achieved in the HAZ zone and in between lower to medium value in TMAZ adjacent to HAZ. The below tables 4.1, 4.2, 4.3 show the respective square, hexagonal and octagonal pin profile tool weldment micro hardness values with various zones. The below figure 4.1, 4.12, 4.3 depicts Vickers micro hardness values graphical presentation of the different pin profile tool weldments of square pin, hexagonal Pin and octagonal pin profile tool.

Table 4.1 Microhardness Values of the various weld zones using Square Pin Profile Tool

Experiment No.	Microhardness in BM -AS	Microhardness in HAZ-AS	Microhardness in TMAZ-AS	Microhardness in NZ	Microhardness in TMAZ-RS	Microhardness in HAZ-RS	Microhardness in BM -RS
1	120	93.57	86.33	87.8	87.03	93.13	121.5
2	119.83	98.63	85.8	94.6	83.87	93.4	117.93
3	118.87	97.93	88.1	91.63	87.38	96.37	118.6
4	121.6	96.17	86.93	86.27	85.07	96.17	127.3
5	118.43	93.63	86.53	92.97	85.53	96.13	117.5
6	119.63	98.23	87.6	94.03	87.5	97.5	116.97
7	124.37	97.63	88.53	93.83	89.6	97.17	117
8	118.8	93.47	87.47	93.63	85.63	92.4	119.57
9	120.3	97.93	87.87	93.6	86	95.4	119.1

The previous table 4.1 depicts the Square Pin Profile Tool weldments microhardness values of the various weld zones. The figure 4.1 shows the graphical illustration of microhardness values of the various FS weld zones using square pin profile tool. Its looks like "W" in shape. The after weldment highest micro hardness value of 98.63$H_V$were noticed in the HAZ-AS side of the serial number 2 weldments, the lowest micro hardness value of 83.87 in TMAZ-RS side serial number 2 weldment. The base metal micro hardness values were in between 124.37 H_V to 117H_V.

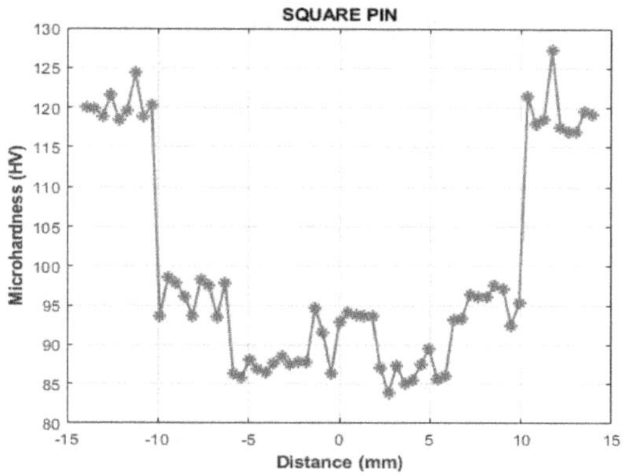

Fig. 4.1 Graphical illustration of Microhardness Values of the various FS weld zones using square pin profile tool.

Table 4.2 Microhardness Values of the various weld zones using Hexagonal Pin Profile Tool.

Experiment No.	**Micro hardness in BM - AS**	**Micro hardness in HAZ-AS**	**Micro hardness in TMAZ-AS**	**Micro Hardness in NZ**	**Micro Hardness in TMAZ - RS**	**Microhardness in HAZ-RS**	**Microhardness in BM - RS**
1	118.83	96.83	90.21	84.17	83.9	95.07	120.07
2	125.93	100.57	88.43	82.97	87.23	100.17	120.43
3	119.3	97.7	86.77	84.33	86.37	98.97	120.97
4	117.83	100.27	87.53	83.43	89.23	99.77	126.92
5	117.77	98.23	84.34	83.57	87.77	89.43	118.57
6	118.73	98.93	87.63	82.7	88.87	94.87	116.47
7	127	94.93	86.23	84.9	84.8	96.17	120.3
8	117.73	94.37	85	85.13	85.27	98.27	120.43
9	121.97	101.47	86.17	84.8	89.37	96.7	118.1

The table 4.2 depicts the Hexagonal Pin Profile Tool weldments microhardness values of the various weld zones. The figure 4.2 shows the graphical illustration of microhardness Values of the various FS weld zones using hexagonal pin profile tool. Its looks like "U" in shape. The after weldment highest micro hardness value of 101.47$H_V$were noticed in the HAZ-AS side of serial number 9 weldment, the lowest micro hardness value of 82.70 in the WNZ of serial number 6 weldment. The base metal micro hardness values were inbetween126.92 Hv to 117.73Hv.

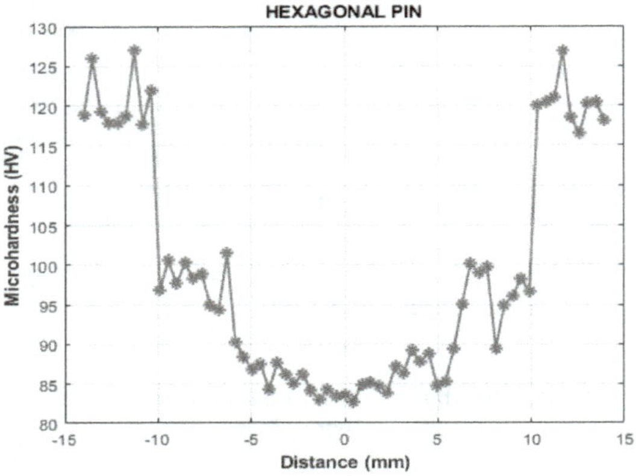

Fig. 4.2 Graphical illustration of Microhardness Values of the various FS weld zones using hexagonal pin profile tool.

Table 4.3 Microhardness Values of the various weld zones using Octagonal Pin Profile Tool.

Experiment No.	Microhardness in BM -AS	Microhardness in HAZ-AS	Microhardness in TMAZ-AS	Microhardness in NZ	Microhardness in TMAZ-RS	Microhardness in HAZ-RS	Micro hardness in BM - RS
1	119.33	95.17	88.97	86.43	88.23	95.17	127.62
2	121.03	95.8	90.47	90.5	89.13	94.57	120.97
3	120.47	95.87	90.03	87.03	88.9	95.47	120.33
4	119.27	94.46	88.84	87.6	93.16	94.14	118.83
5	119.11	96.58	87.25	93.14	91.62	92.69	118.28
6	118.96	98.82	87.84	86.08	87.47	93.86	118.27
7	120.05	97.9	87.03	87.74	90.28	97.77	119.47
8	125.94	93.85	87.61	89.84	90.65	95.58	118.95
9	121.97	97.65	87.56	86.98	86.79	96.79	117.91

The previous table 4.3 depicts the octagonal Pin Profile Tool weldments microhardness values of the various weld zones. The figure 4.3 shows the graphical illustration of microhardness Values of the various FS weld zones using octagonal pin profile tool. It looks like "W" in shape. The after weldment highest micro hardness value of 98.82Hvwere noticed in the HAZ-AS side of serial number 6 weldment, the lowest micro hardness value of 86.08 in WNZ side of serial number 6 weldment. The base metal micro hardness values were inbetween127.62 Hv to 117.91Hv.

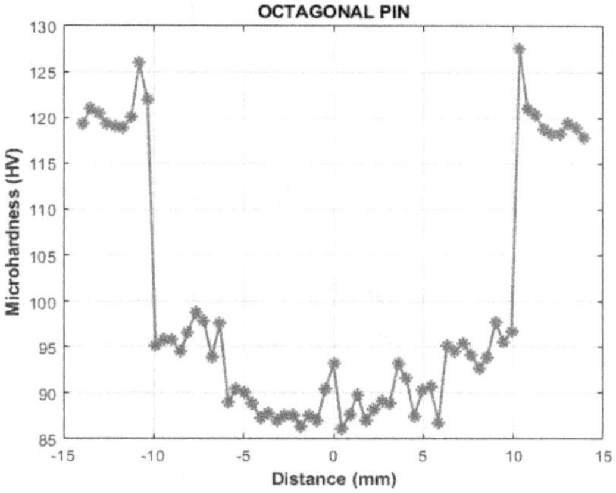

Fig. 4.3 Graphical illustration of Microhardness Values of the various FS weld zones using octagonal pin profile tool.

Among the various profile tool WNZ micro harness values, the highest WNZ micro hardness value of 98.63 HV was observed with square pin profile welded test coupon No:02, and the lowest micro hardness value of 82.70 in the WNZ of serial number 6 weldment of hexagonal profile weldments. The WNZ micro harness values of octagonal profile tool test coupon having moderate among the square and hexagonal pin profile weldments. The octagonal profile WNZ micro harness value of 86.08 HV was observed in serial number 6 weldment. The WNZ area of square pin profile weldments having a harder value than the octagonal pin profile and Hexagonal Pin profile tool weldments, because of the equiaxed grain structure. According to the after-weld hardness assessment, the FSW method dramatically reduces the hardness of the metal in the weld region, around 22 to 31%. This is due to the significant amount of heat input that causes precipitate disintegration, loss of cold work hardening, and elevated temperatures during FSW. The hardness is influenced by the type of tool employed during the friction stir welding. The variation in hardness, in case of welds made by the previous mentioned three different pin profile tools, clearly indicates the softening in the TMAZ and nugget zones due to the exposure to the severe heat input during welding. The hardness of the base material is measured as 117 Hv to 126.92Hv.

A softening region having lower hardness value than the BM exist in all the joints welded at different welding speeds. The hardness of the TMAZ and the HAZ increases with increasing the welding speed, leading to a gradual narrowing of the softening region. This is attributed to the weakening of precipitate deterioration in the two regions as the welding speed increases. Similar TMAZ softening was reported by Liu et al. (2011b), Zhang et al. (2012) and Reddy et al. (2009). Relatively lower hardness values were obtained in all zones of the weld joint made by the conical threaded tool compared to those produced using the hybrid tool.

4.3 Tensile Properties examination of FS weldments AA2219

To assess the Friction, stir weld joint strength, tensile specimens from the welds were cut in a transverse manner across the weldments. The test coupons were sliced according to the American Society for Testing of Materials guidelines of ASTM E8M-04 using wire cut EDM. The tensile test specimen was shown in the below fig. 4.4, The Tensile Testing was carried out in Tinius Olsen Tensile Testing machine (Model H50KL)

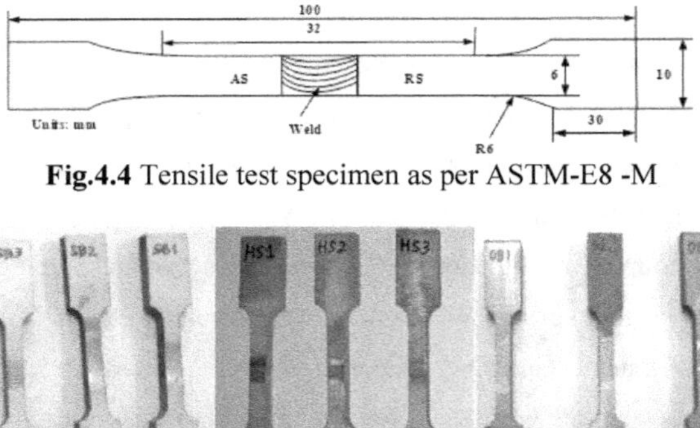

Fig.4.4 Tensile test specimen as per ASTM-E8 -M

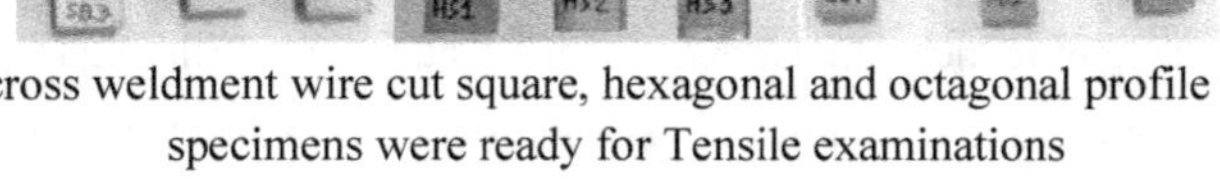

Fig.4.5 Across weldment wire cut square, hexagonal and octagonal profile tool tensile specimens were ready for Tensile examinations

The figure 4.5 shows the across weldment wire cut square, hexagonal and octagonal profile tool tensile specimen ready for Tensile examinations. The experimentally evaluated transverse tensile properties of square, hexagonal an octagonal pin profile welded AA 2219 aluminium alloy joints are presented in the Table 4.4, 4.5, 4.6. The unwelded parent metal shows a yield strength and ultimate tensile strength of 390 and 408 MPa, respectively.

4.3.1 Tensile Properties examination of square pin profile tool weldments.

The experimentally evaluated transverse tensile properties of square pin profile tool are presented in the below table 4.4, The below figure 4.6 shows the graphical illustration of Ultimate Tensile strength with % elongation and weld joint efficiency of square pin profile Tool weldments with error values.

Table 4.4 Ultimate Tensile strength with % elongation and weld joint efficiency of Square pin profile Tool weldments.

Experiment No.	Reference Tool Number	Spindle speed in RPM	Table Travel Speed in mm/min	Tool Tilt Angle in degrees	Ultimate Tensile Stress in Mpa	Total Elongation %	Actual joint Efficiency in % (RM 408 Mpa)
1	SB1	1000	20	2.5	271	13.6	66.42
2	SB2	1000	25	3	293	13.2	71.81
3	SB3	1000	30	3.5	257	7.7	62.99
4	SB4	1250	20	3	249	2.24	61.02
5	SB5	1250	25	3.5	276	11.1	67.64
6	SB6	1250	30	2.5	249	5.98	61.02
7	SB7	1500	20	3.5	214	5.47	52.45
8	SB8	1500	25	2.5	225	4.24	55.14
9	SB9	1500	30	3	209	1.917	51.22

Ultimate Tensile strength with % elongation and weld joint efficiency of square pin profile Tool weldments

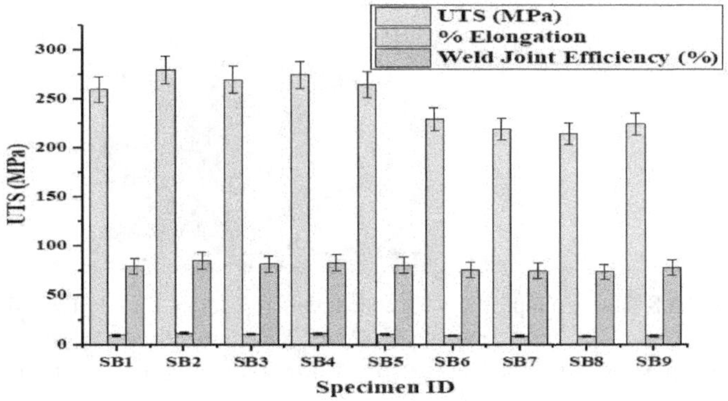

Fig. 4.6 Graphical illustration of Ultimate Tensile strength with % elongation and weld joint efficiency of square pin profile Tool weldments.

The experimentally evaluated transverse tensile properties of square pin profile welded AA 2219 aluminium alloy joints that is having comparatively higher ultimate tensile strength, moderate ultimate tensile strength and lower ultimate tensile strength achieved with respective to their input parameters of square pin profile weldments serial number 2, serial number 6 and serial number 9 are presented respectively in the subsequent below paragraphs.

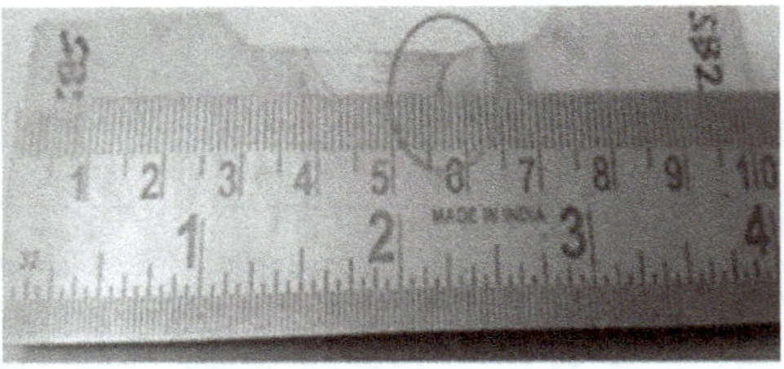

Fig. 4.7 Square pin profile weld specimen Sno.2, after Tensile Test with ductile fracture in WNZ -TMAZ transition area

The previous Fig. 4.7 shows the square pin profile weld specimen Sno.2, after Tensile Test with ductile fracture in WNZ -TMAZ transition area. The serial number 2, FS welding was performed with the tool spindle speed rotation of 1000 rpm with 3^0 tool tilt angle and 25 mm per min. table traverse speed. It yields highest Ultimate Tensile strength of 293 Mpa. with the highest joint efficiency of 71.81%. Due to the comparatively slower tool rotation speed along with a moderate tool tilt angle and with a moderate tool traverse speed influences higher bead width weldment profile material consolidation along with uniformly distributed fine $CuAl_2$ particles joint chemistry with the parent metal.

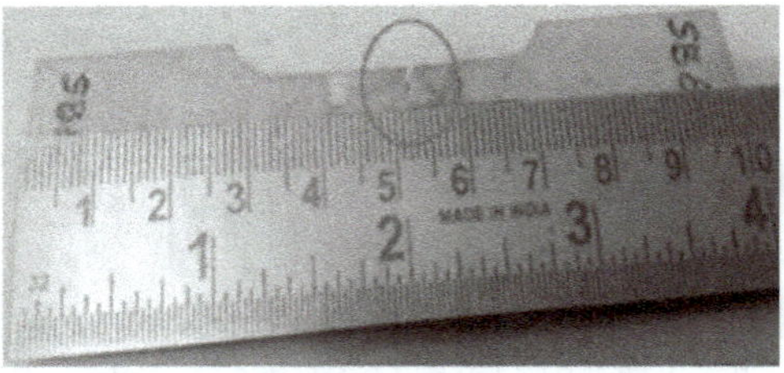

Fig.4.8 Square pin profile weld specimen Sno.6, after Tensile Test with ductile fracture in WNZ -TMAZ transition area

The previous Fig. 4.8 shows the square pin profile weld specimen serial number 6, FS welding was performed with the tool spindle speed rotation of 1250 rpm with 3^0 tool tilt angle and 30 mm per min. table traverse speed. It yields moderate Ultimate Tensile strength of 249 Mpa. with joint efficiency of 61.02%. Due to the higher tool rotation speed with faster travel speed with wider tool tilt angle by the comparatively moderate bead width weldment profile material consolidation with inter related micro chemistry.

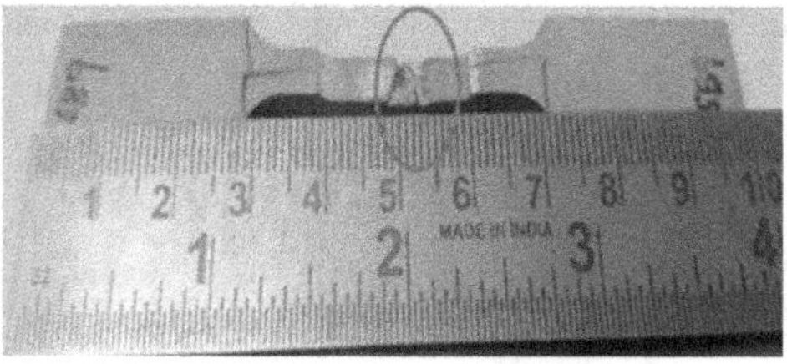

Fig.4.9 Square pin profile weld specimen Sno.9, after Tensile Test with ductile fracture in WNZ -TMAZ transition area

The Figure 4.9 shows the square pin profile weld specimen serial number 9. FS welding was performed with the tool spindle speed rotation of 1500 rpm with 3^0 tool tilt angle and 30 mm per min. table traverse speed yields lower Ultimate Tensile strength of 209 Mpa. with joint efficiency of 51.22%.Due to the higher tool rotation speed with faster travel speed with wider tool tilt angle by the comparatively smaller bead width weldment profile material consolidation interrelated micro chemistry

4.3.2 Tensile Properties examination of hexagonal pin profile tool weldments.

The experimentally evaluated transverse tensile properties of hexagonal pin profile tool were presented in the below table 4.5, The figure 4.10 shows the graphical illustration of Ultimate Tensile strength with % elongation and weld joint efficiency of hexagonal pin profile Tool weldments with error values.

Table 4.5 Ultimate Tensile strength with % elongation and weld joint efficiency of hexagonal pin profile Tool weldments.

Experiment No.	Reference Tool Number	Spindle speed in RPM	Table Travel Speed in mm/min	Tool Tilt Angle in degrees	Ultimate Tensile Stress in Mpa	Total Elongation %	Actual joint Efficiency in % (RM 408 Mpa)
1	HB1	1000	20	2.5	210	5.45	51.47
2	HB2	1000	25	3	258	7.18	63.23
3	HB3	1000	30	3.5	228	5.28	55.88
4	HB4	1250	20	3	201	1.731	49.26
5	HB5	1250	25	3.5	239	7.22	58.57
6	HB6	1250	30	2.5	207	4.42	50.73
7	HB7	1500	20	3.5	202	5.91	49.5
8	HB8	1500	25	2.5	213	5.82	52.2
9	HB9	1500	30	3	198	6.68	48.52

Ultimate Tensile strength with % elongation and weld joint efficiency of Hexagonal pin profile Tool weldments

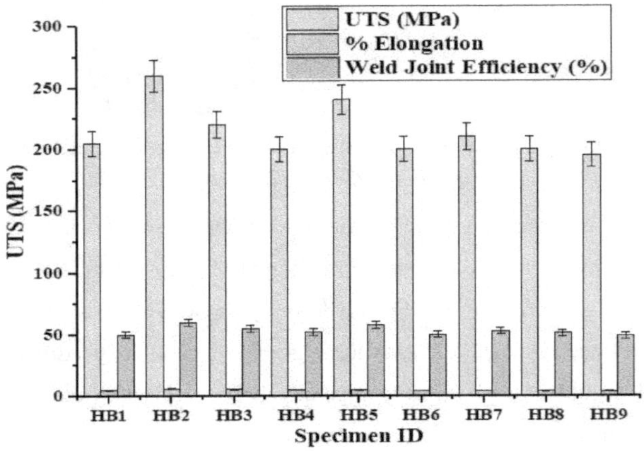

Fig.4.10 Graphical illustration of Ultimate Tensile strength with % elongation and weld joint efficiency of hexagonal pin profile Tool weldments

The Figure 4.10 shows the hexagonal pin profile tool weld specimens Ultimate Tensile strength with % elongation and weld joint efficiency with error values. The experimentally evaluated transverse tensile properties of hexagonal pin profile welded AA 2219 aluminium alloy joints that is having comparatively higher ultimate tensile strength, moderate ultimate tensile strength and lower ultimate tensile strength achieved with respective to their input parameters of hexagonal pin profile weldments serial number 2, serial number 6 and serial number 9 were presented respectively in the subsequent below paragraphs.

Fig.4.11 Hexagonal pin profile weld specimen Sno.2, after Tensile Test with ductile fracture in WNZ -TMAZ transition area

The Figure 4.11 shows the hexagonal pin profile weld specimen serial number 2, FS welding was performed with the tool spindle speed rotation of 1000 rpm with 3^0 tool tilt angle and 25 mm per min. table traverse speed yields Ultimate Tensile strength of 258 Mpa.

with joint efficiency of 63.23%. As per the joint no.2 by the slower tool rotation speed with moderate travel speed with moderate tool tilt angle the comparatively broader bead width weld weldment profile material consolidation inter related micro chemistry.

Fig.4.12 Hexagonal pin profile weld specimen Sno.6, after Tensile Test with ductile fracture in WNZ -TMAZ transition area

The Figure 4.12 shows the hexagonal pin profile weld specimen serial number 6. FS welding was performed with the tool spindle speed rotation of 1250 rpm with 2.5^0 tool tilt angle and 30 mm per min. table traverse speed yields moderate Ultimate Tensile strength of 207 Mpa. with joint efficiency of 50.73%.As per the joint no.6 by the moderate tool rotation speed with faster travel speed with moderate tool tilt angle by the comparatively moderate bead width weldment profile material consolidation interrelated micro chemistry.

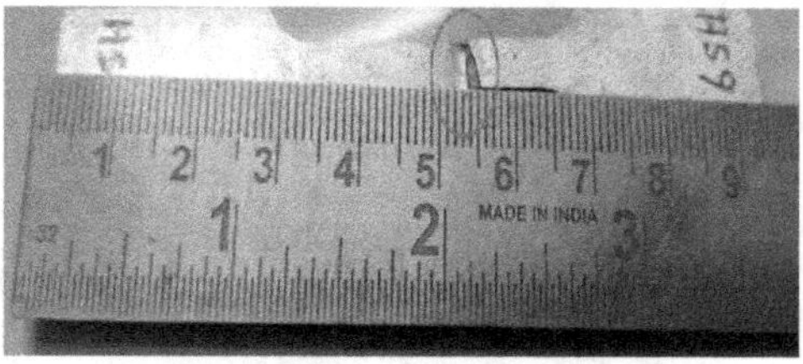

Fig.4.13 Hexagonal pin profile weld specimen Sno.9, after Tensile Test with ductile fracture in WNZ -TMAZ transition area

The Figure 4.13 shows the hexagonal pin profile weld specimen serial number 9. FS welding was performed with the tool spindle speed rotation of 1500 rpm with 3^0 tool tilt angle and 30 mm per min. table traverse speed yields lowest Ultimate Tensile strength of 198 Mpa. with joint efficiency of 48.52%. As per the joint no.9 by the higher tool rotation speed with faster travel speed with wider tool tilt angle by which the comparatively smaller bead width weldment profile material consolidation interrelated micro chemistry.

4.3.3 Tensile Properties examination of octagonal pin profile tool weldments.

The experimentally evaluated transverse tensile properties of octagonal pin profile tool were presented in the below table 4.6, The below figure 4.14 shows the graphical illustration of Ultimate Tensile strength with % elongation and weld joint efficiency with error values of octagonal pin profile Tool weldments.

Table 4.6 Ultimate Tensile strength with % elongation and weld joint efficiency of octagonal pin profile Tool weldments

Experiment No.	Reference Tool Number	Spindle speed in RPM	Table Travel Speed in mm/min	Tool Tilt Angle in degrees	Ultimate Tensile Stress in Mpa	Total Elongation %	Actual joint Efficiency in % (RM 408 MPa)
1	OB1	1000	20	2.5	241	9.79	59.06
2	OB2	1000	25	3	256	9.29	62.74
3	OB3	1000	30	3.5	237	5.41	58.08
4	OB4	1250	20	3	231	6.68	56.61
5	OB5	1250	25	3.5	234	5.98	57.35
6	OB6	1250	30	2.5	241	7.7	59.06
7	OB7	1500	20	3.5	228	7.18	55.88
8	OB8	1500	25	2.5	220	4.24	53.92
9	OB9	1500	30	3	232	2.24	56.86

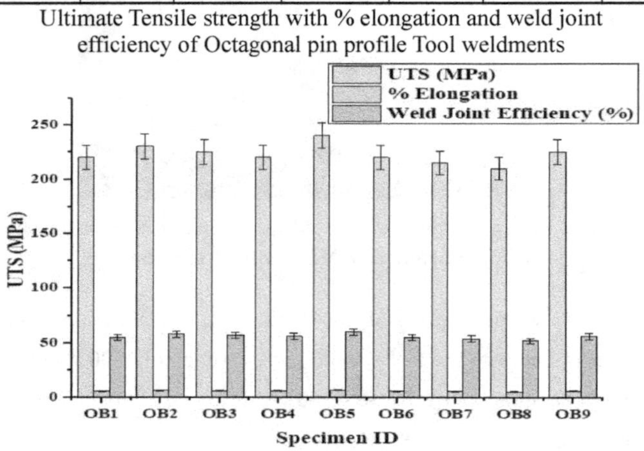

Fig.4.14 Graphical illustration of Ultimate Tensile strength with % elongation and weld joint efficiency of octagonal pin profile Tool weldments

The experimentally evaluated transverse tensile properties of octagonal pin profile welded AA 2219 aluminium alloy joints that is having comparatively higher ultimate tensile strength, moderate ultimate tensile strength and lower ultimate tensile strength achieved with respective to their input parameters of octagonal pin profile weldments serial number 2, serial number 5 and serial number 9 were presented respectively in the subsequent below paragraphs.

Fig.4.15 Octagonal pin profile weld specimen Sno.2, after Tensile Test with ductile fracture in WNZ -TMAZ transition area

The previous Fig. 4.15 shows the octagonal pin profile weld specimen serial number 2, FS welding was performed with the tool spindle speed rotation of 1000 rpm with 3^0 tool tilt angle and 25 mm per min. table traverse speed yields higher Ultimate Tensile strength of 256 Mpa. with joint efficiency of 62.74%. As per the joint no.2 slower tool rotation speed with moderate tool traverse speed and with moderate tool tilt angle by the higher bead width weldment profile with uniformly distributed fine $CuAl_2$ particles joint chemistry with the parent metal.

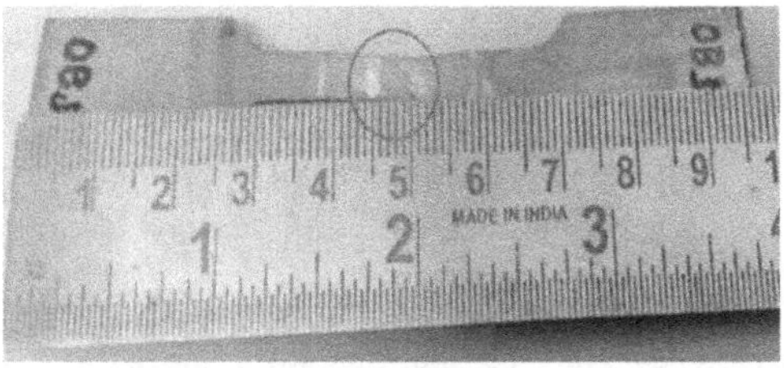

Fig.4.16 Octagonal pin profile weld specimen Sno.5, after Tensile Test with ductile fracture in WNZ -TMAZ transition area

The previous Fig. 4.16 shows the octagonal pin profile weld specimen serial number 5. FS welding was performed with the tool spindle speed rotation of 1250 rpm with 3.5^0 tool tilt angle and 25 mm per min. table traverse speed yields moderate Ultimate Tensile strength of 234 Mpa. with joint efficiency of 57.35%. As per the joint no. 5 moderate bead width weldment profile by the moderate tool rotation speed with moderate tool traverse speed and with higher tool tilt angle along with uniformly distributed fine $CuAl_2$ particles joint chemistry with the parent metal.

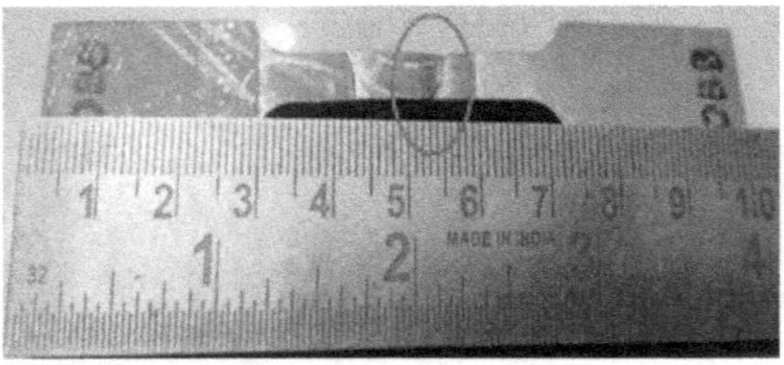

Fig.4.17 Octagonal pin profile weld specimen Sno.8, after Tensile Test with ductile fracture in WNZ -TMAZ transition area

The Figure 4.17 shows the octagonal pin profile weld specimen serial number 8, FS welding was performed with the tool spindle speed rotation of 1500 rpm with 2.5^0 tool tilt angle and 25 mm per min. table traverse speed yields lower Ultimate Tensile strength of 220 Mpa. with joint efficiency of 53.92%. As per the joint no.8 , lower bead width by the higher tool rotation speed with tool moderate traverse speed and with moderate tool tilt angle along with uniformly distributed fine CuAl2 particles joint chemistry with the parent metal.

From the post tensile specimens in most of the cases found ductile fracture was found near TMAZ to HAZ adjoining areas of AS or RS side. The average ultimate TS of Hexagonal pin profile welded joints was found 217.33 Mpa. The highest Tensile value was achieved with HP profile welded with test coupon no:2 with a value of 258 Mpa with joint efficiency of 63.23%. The average ultimate TS of Octagon pin was found 243 Mpa. The highest Tensile value was achieved with OP profile welded with test coupon no:2 with a value of 256 Mpa with joint efficiency of 62.74%. The average ultimate TS of SP profile WJ was found 249.77 Mpa. The highest Tensile value was achieved with SP profile welded with test coupon no:02 with a value of 293 Mpa. with joint efficiency of 71.81%. From three Pin Profile, the square pin tool has the highest UTS value of 293 Mpa and lower most tensile value was achieved with HP Test coupon no:9 with a tensile value of 198 Mpa. Which one having micro void defect found on the advancing side WNZ bottom portion may lead to poor joint strength.

From the result it is evident that the particular Tool Pin Profile, welding speed and Tool Tilt Angle have significant impact on Microstructure, while the micro structure has significant impact on Tensile Strength. The experimental results of weld joints show that the pin geometry has significant impact with a vital role in achieving the joint strength. The flat faces with the tool rotation, increase spinning flow of material with frictional heating and the annealing action during FSW causes a very fine grain refinement of Micro Structure. The effect SP with high eccentricity influencing the micro structure of ultrafine equiaxed grains by which the highest tensile value is obtained.

The reduction of heat by an increase in welding speed causes reduced precipitate coarsening in the HAZ, thus leading to a moderately increased joint strength with the fracture location shifting towards the TMAZ from the HAZ, as reported by Zhang et al. (2011) and Liu et al. (2011a). Increasing the welding speed beyond 200 mm/min might result in a top groove defect because of the difference between the material flow speed and the tool travel speed (Li and Liu, 2013). The three different pin profile of square, hexagonal and octagonal welded joints, the hexagonal pin profile welded joints show the lowest yield strength and tensile strength of 220 and 242 MPa, respectively (Liu, J. et.al 2022). This suggests that there is a 50% reduction in strength values due to hexagonal pin profile welding. Square pin profile joints show the highest yield strength and tensile strength of 305 and 342 MPa, respectively. The structure consists of pancake shaped larger grains. The elongated grain morphology appeared along the rolling direction of the base plate with average grain size of 63 mm. The grains with low strains were aligned 5^0 to 50^0 with respect to the rolling direction. The base material exhibited an Ultimate Tensile Strength of 442 MPa and Yield Strength of 335 MPa. This is mainly the Al-Cu composition with the T87 age hardening effect (Ahmed et.al 2017)

4.4 Macrostructural Examination of FSW AA2219-T87

FSW bead macrostructure analysis typically focuses on the appearance and distribution of material across the welded region, and it helps to evaluate the overall integrity of the weld, detect defects. The figure 4.18 depicts the AA2219- T87 Macro structure images of various pin profile tool weldments and the various zones of the friction stir welding.

Fig. 4.18 Macro images of FSW joints with various weld zones

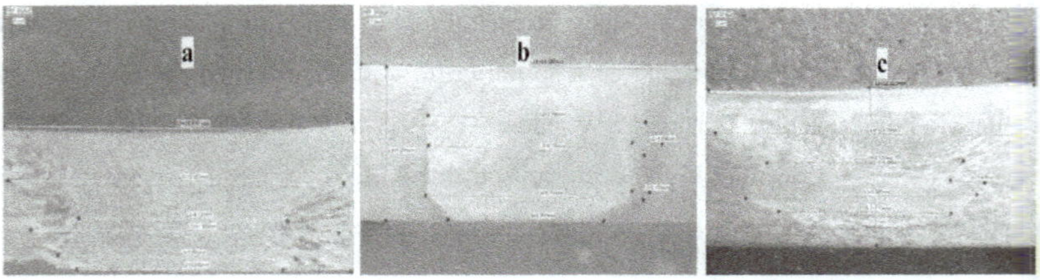

Fig. 4.19 Macrostructure images of AA2219-T87 - (a) Square pin profile (b) Hexagonal pin profile (c) Octagonal pin profile tool weldments

From the macro structural analysis, the after-welding bead width variations of the top and bottom was varied according to the different tool pin profiles of square, hexagonal and octagonal. The figure 4.19(a) shows the macrostructure of AA2219-T87 – Square pin profile weldment SNo.2, figure 4.19(b) shows the macrostructure of AA2219-T87 – hexagonal pin profile weldment SNo.2 and the figure 4.19(c) shows the macrostructure of AA2219-T87 – Octagonal pin profile weldment SNo.2. The bead width of square pin profile topmost layer was 18.970 mm and on bottom most layer was 8.040 mm. The bead width of hexagonal pin profile on top most layer was 18.674 mm and on bottom most layer was 5.827 mm. The bead width of octagonal pin profile on topmost layer was 18.628 mm and on bottom most layer was 7.227 mm. The Stir zone looks like a bowl in hexagonal pin profile welding. The Stir zone looks like an inverted half pot in square and octagonal pin profile welding. The upper surface experiences more frictional heat and drastic plastic flow due to the contact with tool's shoulder, which results in a wider upper surface. The pin tool extends into the material thickness and directly influences material mixing and grain structure. It is essential for achieving proper bonding between the faying surfaces. The bottom portion of the joint material only exposed by the pin profile diameter or size for a shorter width only.

In Friction Stir Welding (FSW), the swept volume of the tool depends on its shape and the path it follows during the welding process. The dynamic volume swept by the tool is the space it occupies as it rotates and moves along the weld line. For different tool profiles like square, hexagonal, and octagonal, the swept volume is calculated based on their cross-sectional geometry and their rotation during welding. The dynamic volume is the actual volume of material stirred or affected dynamically by the tool as it travels along the weld line. It depends on the tool's linear velocity (travel speed) and rotational speed.The swept volume and dynamic volume are important in friction stir welding (FSW) to understand how the pin profile impacts metal flow, stirring, and heat generation. The improvements in joint properties have been associated with the pulsating action of the pin while stirring, which is related to the polygonal corners and the eccentricity ratio, i.e. the ratio of the dynamic volume swept by the tool to the static volume of the tool

Table 4.7 Various FSW Tool pin profile with their pulse rate and eccentricity ratio

Pin profile	Top view of pin profile	Pulse rate at 1000 rpm (pulse/s)	Eccentricity ratio
Straight cylindrical		-	$\pi r^2/\pi r^2 = 1$
Straight triangle		50	$4\pi/(3\sqrt{3}) = 2.42$
Straight square		67	$\pi/2 = 1.57$
Straight hexagon		100	$2\pi/(3\sqrt{3}) = 1.21$
Straight octagon		133	$\pi/(1.17(1+\sqrt{2})) = 1.11$

It is clear from the Table 4.7 that pulse rate increases with the number of flat faces on the pin profile. Since a hexagonal pin profile will produce 6 pulses per rotation, which is double the pulse rate produced by a triangle pin profile, its pulse rate also doubles. Additional frictional heat is generated owing to the pulsating stirring action, which offers resistance to the regular flow of plasticized material (R. Palanivel 2012). In addition, works by Gratecap et al. (2012) had shown that triangle and square tool pin profiles produce material flow patterns which feature three or four segmented layers corresponding to the three or four vertices on the tool pin (Gratecap et al.2012). Thus more plasticized material and enhanced material flow can be predicted.

4.5. Microstructural Analysis of FSW AA2219-T87

The friction stir weld aluminium alloys consists of different zones. The macrostructure of an FSW bead consists of several distinct regions formed by different thermal and mechanical conditions during the welding process. The soften region composed of the weld nugget zone (WNZ), the thermal mechanically affected zone (TMAZ) and the heat affected zone (HAZ) tends to be created by FSW thermal cycles due to the coarsening or dissolution strengthening of precipitates. Fonda RW et al. (2004), Mishra RS et al. (2005), Simar A et.al (2008), Starink MJ (2008).

4.5.1 Microstructural Characterization of Base Metal Area (BM Zone)

The Micro Structure of AA2219 is typically composed of aluminium matrix and reinforcing particles of copper, magnesium, and manganese. These particles can form clusters, and their distribution and size may affect the alloy's strength and ductility. The base material (BM) acts as the reference material for analysing the influence of FSW on AA2219. It shows an essential element with its original microstructure and mechanical properties. Comparing the features of the base metal with those of the separate welded zones provides an in-depth examination of ways the FSW technique affects the material. It also provides the existence of precipitate particles within distinct sections of the base material. These microstructural particles contribute to the overall heterogeneity of the alloy, altering its mechanical and thermal characteristics.

The base material consists of typically elongated grains measuring approximately to 2–3 mm with θ ($CuAl_2$) particles spread across the whole matrix. The composition of the AA2219 yields a microstructure composed primarily of the saturated aluminium matrix plus a small amount of excess equilibrium (Al_2Cu) phase. The T87 temper used in this study refers to a heat treatment that artificially ages the Cu rich precipitates in the Alpha-matrix through a well-accepted sequence. On microscopic analysis of base metal, it reveals the grains of larger pancaked shape. The elongated pancake shaped grain morphology appeared along the rolling direction of the base plate with average grain size of 63 μm.

AA2219 has Cu concentration above the maximum solid solubility of 5.35%, Al2Cu is in excess equilibrium phase is usually present in the starting rolled plate material. Depending on the temperature history of the FSW, the presence of the equilibrium phase has been reported to change (Cao and Kou, 2005).

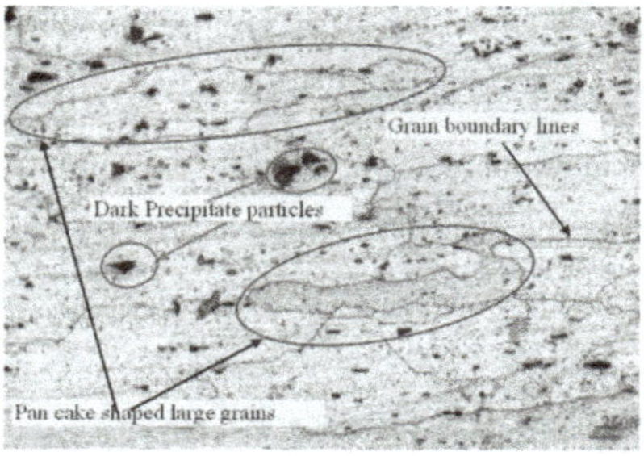

Fig.4.20 AA2219-T87 Parent metal microstructure

The figure 4.20 shows the parent material microstructural images, the structure consisted of pancake shaped larger grains with their grain boundaries with dark precipitates dispersed arrangements.

4.5.2 Microstructural Characterization of Weld Nugget zone (WNZ/SZ/ DRZ)

The WNZ is formed underneath of the pin where the middle of firm joint occurs. Due to the extreme plastic deformation and extrusion action of tool spinning at a high temperature, this nugget zone gets exposed into a recrystallized microstructure of very tiny sized equiaxed grains.

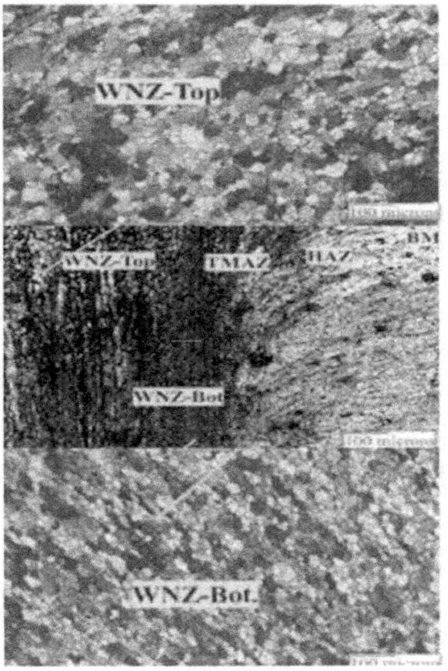

Fig. 4.21 Square pin profile tool after weld microstructural reveals layers of coarse and fine grains in WNZ-Top and Bottom respectively.

The Figure 4.21 shows the after-weld micrograph of a square tool pin profile weldment. The nugget zone of metal joint appears to be different grain size at the top and bottom of the WNZ due to the difference in thermal heating experience of the top and bottom portions of the weld nugget zone. while the grain size within the DRZ is uniform in some alloys, others reveal layers of course and fine grains. It is proposed that this feature is caused by cyclic or layered flow around the pin, which causes high and low rates of deformation. Note, a cylindrical tool has no features to produce this periodic behaviour. The higher the deformation the finer the resulting grain size from dynamic recrystallisation. The top of the nugget zone is exposed to a higher degree of heating and the larger centrifugal force due to the close contact with the tool shoulder, which further results in grain growth after dynamic recrystallization. For the alloy 2219, Cu is a principal alloying element, and the primary strengthening precipitates are the metastable θ and θ' phases [Onol A.T., 2010). on their study, the measured diameter of these precipitates ranged from 25 to 200 nm. This apparent decrease of grain size is due to the heat sink effect of the steel backing plate, which draws away the heat, making less and less available for grain growth. The measured grain size for the base material in the traverse direction was 25 μm an average value of 2.5 μm was observed for the weld nugget. The bottom of the nugget zone is in contact with the backing plate that acts as a heat sink. The bottom zone experiences shorter weld thermal cycles and lower peak temperatures, which slows the grain growth, resulting in smaller grain size at the bottom of nugget zone. These observations are in line with the findings of Xu et al. (2013).

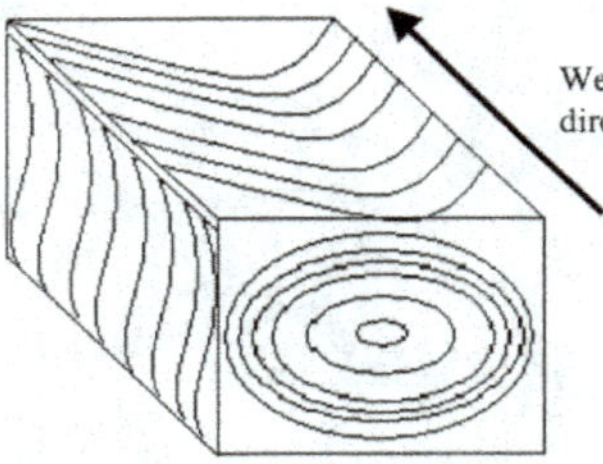

Fig. 4.22 Schematic view of the extruded layers of weld material in semi-cylindrical shape ring structure by the FSW tool rotation with forward motion (Ref. Krishnan 2002)

In FSW microstructures, the Dynamically Recrystallised Zone (DRZ) are far from continuous and show features which are periodic with the tool rotation. This has traditionally been labelled as the 'onion ring' and is clearly seen in the planar macro section in the Figure 4.22. According to stir theory, each rotation of the tool during FSW causes layers of material that are semi-cylinder shapes to be extruded, which when examined from a cross-sectional slice, exhibit the well-known onion ring structure in the weld direction. The metal has undergone significant deformation and has been fused together to form a strong joint. The formation of onion ring is due to the geometric effect that cylindrical sheets of material are extruded during each rotation of the tool and the cutting through the section of the material produces an apparent 'Onion Rings'.

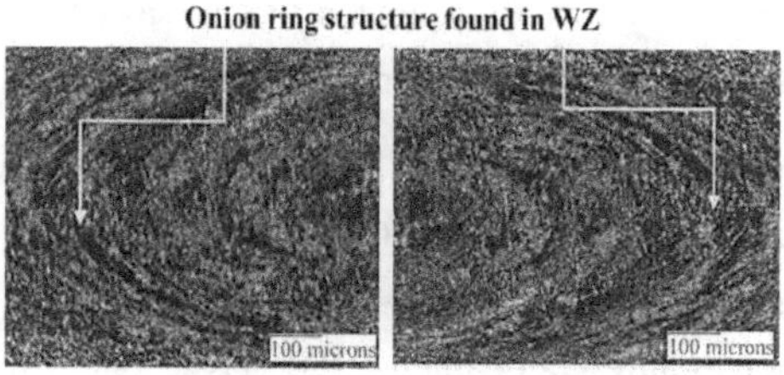

Figure 4.23: WNZ Onion Ring formation from the square pin profile tool

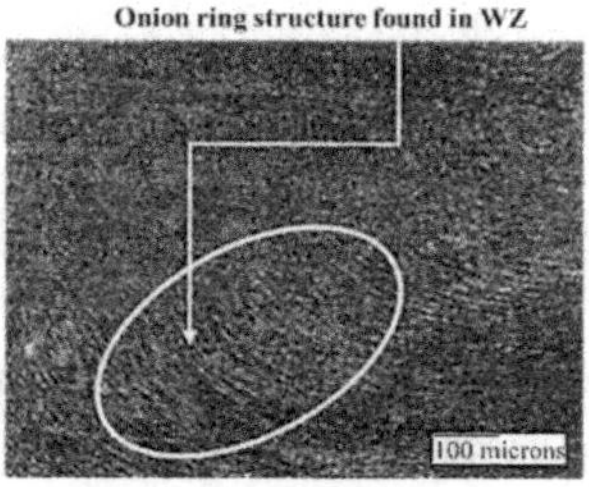

Figure 4.24 WNZ Onion Ring formation from the octagonal pin profile tool

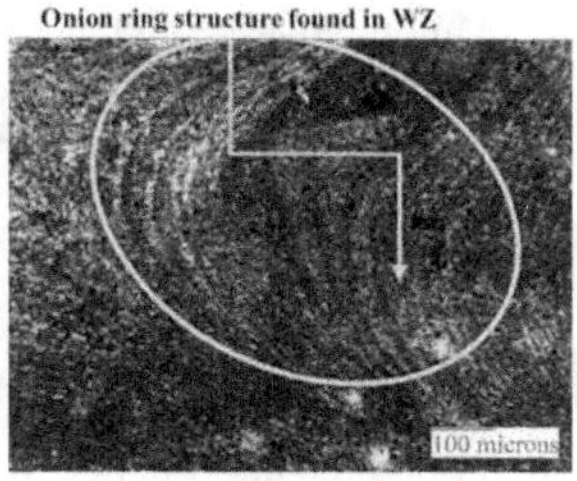

Figure 4.25 WNZ Onion Ring formation from the hexagonal pin profile tool

The figures 4.23, 4.24 and 4.25 show the Onion Ring formation in the weld nugget by the extruding spinning action of square pin, the octagonal pin and hexagonal pin profile tool while welding respectively, Among all the above three tool weldments, the WNZ of the octagonal pin profile grains were ultra fine equiaxed, comparatively smaller in size than square and hexagonal pin profile tool weldments microstructure. The square pin profile weldment WNZ grains were comparatively bigger in size and the hexagonal profile weldment WNZ grains were in between the square and octagonal pin profile weldments. The figure 4.25 shows the weld nugget formation of a hexagonal pin profile tool the grains which were very fine equiaxed in-between the square and octagonal pin profile WNZ microstructural grain size. The onion ring formation was associated with the forward motion of the tool in one revolution. But it did not provide any explanation about the onion ring formation, but this may be affecting the properties of the weld (Threadgill 1999).

The spacing of the markings has been found to be equal to the forward motion of the tool in one rotation. The tool extruded onion rings of semicylinders would appear like cut vegetable onion rings with ring spacing being wider at the centre and narrower towards the edge. For the dissimilar-alloy configuration, the classic onion ring's structure has a different morphology to be of asymmetric vortex-shaped. The asymmetric vortex-shaped region was slightly towards the right of the weld centreline and coincides with the chemical mixing zone. One of the features of the onion rings most often referred to is the decreasing distance between the rings when progressing from the centre towards the periphery (Krishnan 2002).

A clear onion pattern in the stir zone is seen in optical microstructure and SEM micrographs, providing optimum material blending with continuously recrystallized thin equiaxed grains (Rafiei. R et al. 2017). The distance between the onion rings is based on tool travel speed during welding (Muthukumaran, S et al. 2017).

4.5.3 Microstructural characterization of Thermo-Mechanically Affected Zone (TMAZ)

The TMAZ undergoes plastic deformation and partial recrystallization due to high temperatures and intense strain. In TMAZ, it is illustrated that the parent grains tend to be reoriented along the flow pattern around the tool pin due to shear deformation induced by the rotating tool, which is the typical feature of TMAZ. Newly-formed equiaxed grains are obviously observed, indicating that partial dynamic recrystallization (DRX) occurs. A tendency can be seen that from the outside to inside of the TMAZ (i.e. from left to right in Fig. 4.26(a) and (c)), more equiaxed grains are formed. With approaching closer to weld centre, higher temperature and larger locational deformation are achieved, consequently the extent of DRX increases.

The Figure 4.26 (a–e) shows the TMAZ micro structural morphology. It is a transition region between the nugget zone and the HAZ. The TMAZ experiences both stirring-assisted mechanical working and severe weld thermal cycles that result in a heavily deformed grain structure and associated grain growth. It is evident from Figure 4.26(a) and 4.26(c) that, from top to bottom there exists a large variation in the width of the TMAZ throughout the thickness. The TMAZ under shoulder is wider at the shoulder and induces extra heat and centrifugal force during stirring. The TMAZ is narrowest at the bottom of the nugget zone as the deformation and heat generated by the tool pin is comparatively smaller in this zone.

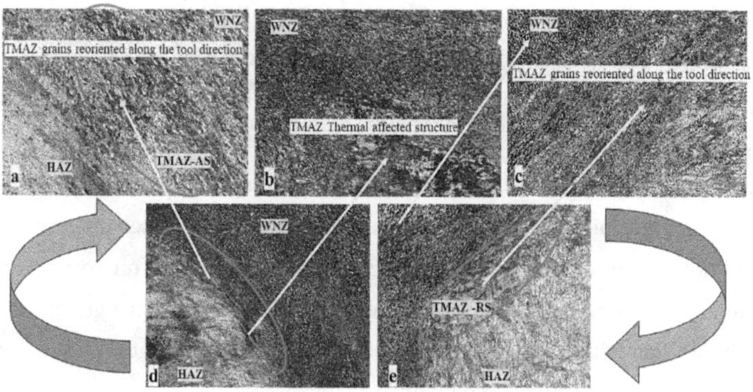

Figure 4.26 TMAZ micro structural morphology with grains reoriented along the tool direction flow pattern (Hexagonal pin profile)

The TMAZ micrographs exhibit coarser grains and so reduced grain boundary strengthening is observed in this region. Thus, the lower hardness of TMAZ is attributed to the low grain boundary strengthening and precipitate hardening. The age-hardenable materials are mainly strengthened due to the presence of fine θ' precipitates. During loading, these precipitates act as the obstacle for the dislocation motion. The coarse stable θ precipitates cannot provide the hindrance effect due to low coherency. From the Hall-Petch relation, it was understood that the hardness or strength decreases with an increase in the grain size (Xu et al. 2012). The Figure 4.26 (a)&(c) clearly shows that the elongated grains were formed at both advancing side and retreating side of the TMAZ during welding, owing to the plastic shear stress along the material flow in the weld zone This was confirmed by the same characteristic material flow in their experiment 2 and 1 (Venkata Rao et al. (2015), Xu et. al (2012)).

The TMAZ had a peak temperature changing from 653^0K to 673^0K (380^0C to 400^0C) during welding, which also made θ" and θ' phases completely redissolve into the matrix. However, θ phase did not redissolve. As a result, the amount of SSS transformed from precipitates was less in the TMAZ than that in the WNZ. Meanwhile, the TMAZ did not undergo dynamic recrystallization to refine grains. (J kang et al. 2016)

4.5.4 Microstructural characterization of Heat Affected Zone (HAZ)

The base metal at both extreme end is depicted in the Figure. 4.27(a) It is far away from the tool rotational zone free of thermal and mechanical effects, and shows obvious elongated grain microstructure without the impact of frictional heating and occurrence of plastic deformation. The HAZ portion is away from the tool rotational zone (Fig. 4.27 b) and only experiences limited heat transfer from TMAZ which suffered much less thermal influence, and hence HAZ exhibits similar grain structures as like the base metal. The HAZ just experiences a thermal cycle heating which is transferred from adjacent TMAZ and does not undergo appreciable plastic deformation. The temperature is comparatively lower which does not influence the microstructural changes. Therefore, the microstructure in the HAZ is similar in appearance of the parent metal micro structure and the grains are coarser in size on either side of the weld nugget, there are no noticeable changes. The coarse grains are observed in the HAZ compared with other regions.

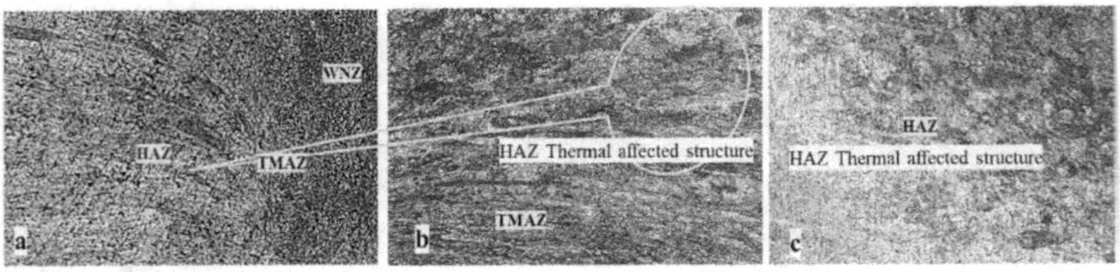

Fig. 4.27 Micro structural morphology of hexagonal pin profile weldment HAZ, with limited thermal influence from TMAZ makes the grains coarser in size as like BM.

The HAZ underwent a wide temperature range of 453^0 K to 653^0 K (180^0 C to 380^0 C), where θ" phase partly to completely redissolved into the matrix, while some θ' phase also redissolved and the rest coarsened. The amount of redissolved θ' phase decreased with increasing distance away from the weld centreline and thus the hardness increased gradually (J kang et.al 2016)

4.5.5 Microstructural characterization of square pin profile pin weldments

The figure 4.28 shows the microstructure of BM, WNZ, TMAZ, HAZ Zones of Square Pin Profile weldment. The microstructural grains of the Square Pin Profile produced joints obtained via optical microscope, SEM are presented in Figure 4.28 (a-f). After FSW, fine equiaxed grains were achieved as a result of dynamic crystallization, and they were significantly smaller than those in the BM. It can be seen that the WNZ of the impulse welds possessed more homogeneous grains distribution, compared to the FSW structure.

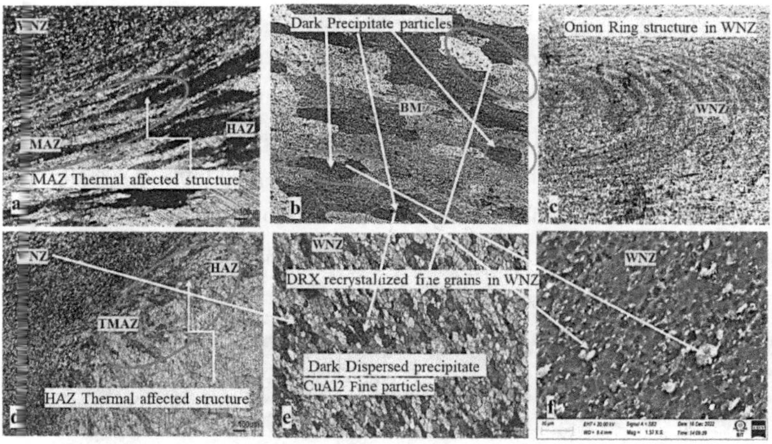

Fig. 4.28 Microstructure of BM, WNZ, TMAZ, HAZ Zones of Square Pin Profile Weldment

Figure 4.28(b) shows BM structure having the inherited coarse intermetallic particles of $CuAl_2$. Figure 4.28(c) shows the onion ring grain structure formation in the WNZ. Figure 4.28(e) and 4.28(f) show the ultrafine parent grains with $CuAl_2$ and intermetallic particles were also visible in the microstructures of the WNZ, indicating their insoluble nature or only partial dissolution, considering slight decrease in their average size. Figure 4.28(f) shows the EDX morphology of the precipitate distribution in the WNZ of the FSW weld. It demonstrated round particles with an average diameter in the range of 100–120 nm which were also observed in the microstructure of the BM Figure 4.27(e) shows Weld nuggets featuring refined grains with dynamic recrystallization, especially with square and hexagonal pin profiles. These changes improve mechanical properties by distributing second-phase θ ($CuAl_2$) particles uniformly. The tool geometry significantly influences grain size and precipitate behaviour (Sahu et al., 2023).

The subsequent paragraphs 4.5.5.1, 4.5.5.2 and 4.5.5.3 of microstructural characterization of square pin profile tool deals about the best, moderate and least weld joint results micro structure morphology in terms of higher tensile strength, higher microhardness and lower corrosion rate from the higher level to lower levels respectively.

4.5.5.1 Microstructural characterization of square pin profile pin-S.No.2 weldments

The figure 4.29 shows the microstructural morphology of sno.2 square pin profile various weldment zones. The figure 4.29(a) shows the TMAZ. The figure 4.29(b) shows the WNZ

The figure 4.29(c) shows the HAZ. On the serial number 2, FS welding was performed with the tool spindle speed rotation of 1000 rpm with 3^0 tool tilt angle and 25 mm per min. table traverse speed which yields highest Ultimate Tensile strength of 293 Mpa. with the highest joint efficiency of 71.81% , as per the slower tool rotation speed along with a moderate tool tilt angle and with a moderate tool traverse speed influences higher bead width along with uniformly distributed fine $CuAl_2$ particles joint chemistry with the parent metal. The same were noticed by the previous researcher. (Sahu et al., 2023)

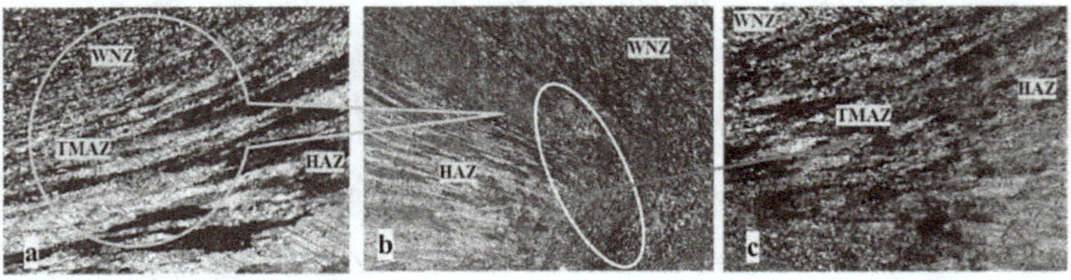

Fig. 4.29 S.No.2 SBT Weld specimen WNZ, TMAZ, HAZ Thermal affected microstructure

4.5.5.2 Microstructural characterization of square pin profile pin-S.No.6 weldments

The figure 4.30 shows the microstructural morphology of sno.6 square pin profile various weldment zones. The figure 4.30(a) show the TMAZ. The figure 4.30(b) shows the WNZ.The figure 4.30(c) shows the HAZ. On the serial number 6, FS welding was performed with the tool spindle speed rotation of 1250 rpm with 3^0 tool tilt angle and 30 mm per min. table traverse speed which yields moderate Ultimate Tensile strength of 249 Mpa. with joint efficiency of 61.02%, as per the comparatively smaller bead width with inter related micro chemistry by the higher tool rotation speed with faster travel speed with wider tool tilt angle.

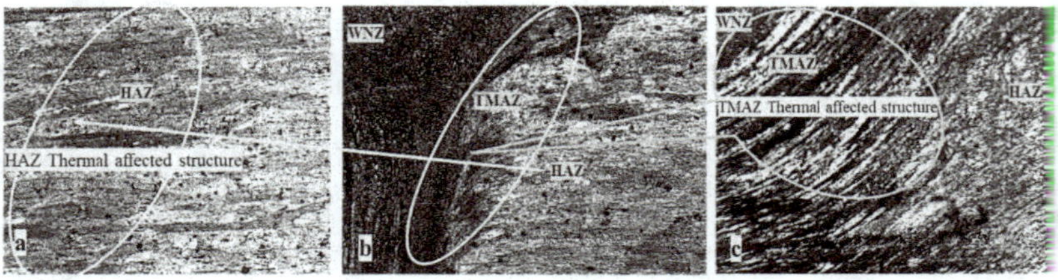

Fig. 4.30 S.No.6 SBT Weld specimen WNZ, TMAZ, HAZ Thermal affected microstructure

4.5.5.3 Microstructural characterization of square pin profile pin-S.No.9 weldments

The figure 4.31 shows the microstructural morphology of SNo.9 square pin profile various weldment zones. The figure 4.31(a) shows the TMAZ. The figure 4.31(b) shows the WNZ. The figure 4.31(c) shows the HAZ. On the serial number 9, FS welding was performed with the tool spindle speed rotation of 1500 rpm with 3^0 tool tilt angle and 30 mm per min. table traverse speed which yields lower Ultimate Tensile strength of 209 Mpa. with joint efficiency of 51.22% , as per the comparatively smaller bead width by the higher tool rotation speed with faster travel speed with wider tool tilt angle weld material consolidation inter related micro chemistry.

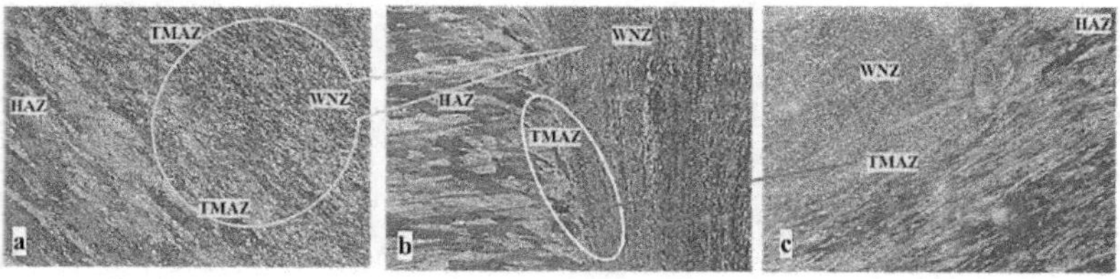

Fig.4.31 Sno.9 SBT Weld specimen WNZ, TMAZ, HAZ Thermal affected microstructure

4.5.6 Microstructural characterization of hexagonal pin profile pin weldments

Microstructures of different weld zone of hexagonal pin profile trial-2 weld joint are presented in the Figure 4.32 (a-e) The Figure 4.32 (e) shows the WNZ portion of DRX recrystallized fine grain structure. The hexagonal pin profile extrusion by local frictional heating and severe plastic strain effects causes dynamic recrystallization underneath the pin portion makes the grains into finer dispersed. During FSW, an increase in the combined effect of frictional heating and plastic deformation takes place from TMAZ to nugget (K.V. Jata et.al 2000). The WNZ grain size in trial-2 was comparatively bigger than the trials 6 and 9, which is shown in the subsequent Figure 4.32 (e). The Figure 4.32 (a) clearly shows that the elongated grains were formed at advancing side of the TMAZ during welding, owing to the plastic shear stress along the material flow in the weld zone. This was confirmed by the same characteristic material flow in experiment 2 and 1 (Venkata Rao et al. (2015), Xu et al (2012)). The subsequent paragraphs 4.5.6.1, 4.5.6.2 and 4.5.6.3 of microstructural characterization of hexagonal pin profile tool deals about the best, moderate and least weld joint result's micro structure morphology in terms of higher tensile strength, higher microhardness and lower corrosion rate from the higher level to lower levels respectively.

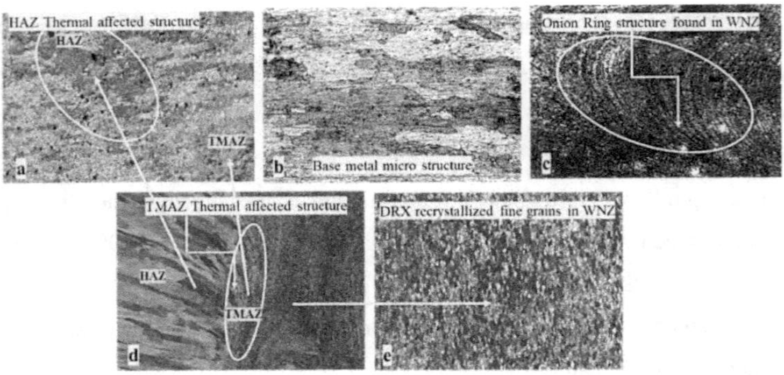

Fig. 4.32. Microstructure of BM, WNZ, TMAZ, HAZ Zones of Hexagonal Pin Profile weldment

4.5.6.1 Microstructural characterization of hexagonal pin profile pin-S.No.2 weldments

The figure 4.33 shows the microstructural morphology of sno.2 hexagonal pin profile various weldment zones. The figure 4.33(a) shows the TMAZ. The figure 4.33(b) shows the WNZ The figure 4.33(c) shows the HAZ.

On the serial number 2, FS welding was performed with the tool spindle speed rotation of 1000 rpm with 3^0 tool tilt angle and 25 mm per min. table traverse speed which yields higher Ultimate Tensile strength of 258 Mpa. with joint efficiency of 63.23% due to the comparatively broader bead width by the slower tool rotation speed with moderate travel speed with moderate tool tilt angle weld material consolidation interrelated micro chemistry.

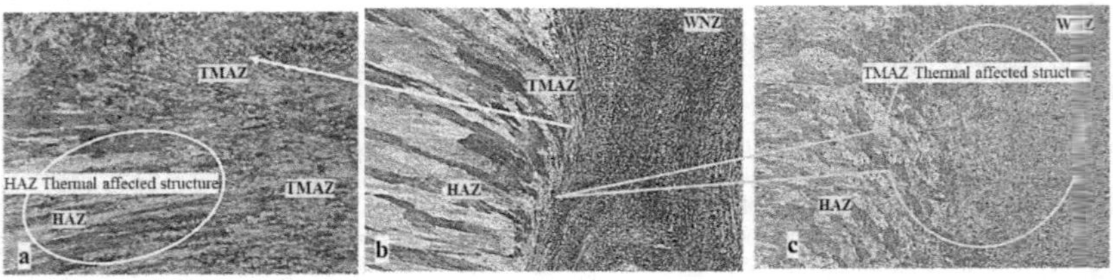

Fig. 4.33. S.No.2 Hex Ball tool WNZ, TMAZ, HAZ Thermal affected microstructure

4.5.6.2 Microstructural characterization of hexagonal pin profile pin-S.No.6 weldments

The figure 4.34 shows the microstructural morphology of S.No.6 hexagonal pin profile various weldment zones. The figure 4.34(a) shows the TMAZ. The figure 4.34(b) shows the WNZ The figure 4.34(c) shows the HAZ. On the serial number 6, FS welding was performed with the tool spindle speed rotation of 1250 rpm with 2.5^0 tool tilt angle and 30 mm per min. table traverse speed which yields moderate Ultimate Tensile strength of 207 Mpa. with joint efficiency of 50.73% due to the comparatively moderate bead width by the faster travel speed with moderate tool tilt angle weld material consolidation inter related micro chemistry.

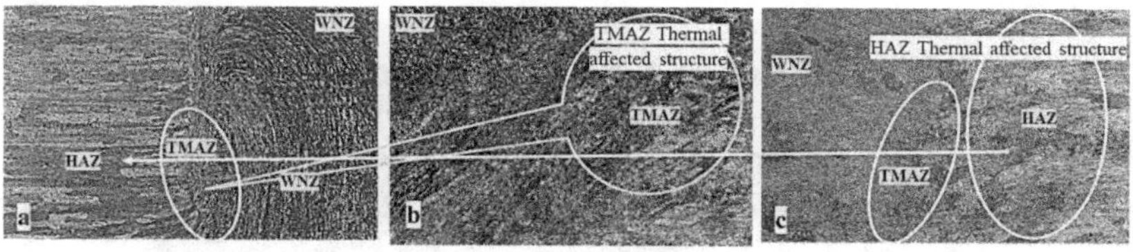

Figure 4.34: S.No.6 Hex Ball tool WNZ, TMAZ, HAZ Thermal affected microstructure

4.5.6.3 Microstructural characterization of hexagonal pin profile pin-S.No.9 weldments

The below figure 4.35 shows the microstructural morphology of S.No.9 hexagonal pin profile various weldment zones. The figure 4.35(a) shows the TMAZ. The figure 4.35(b) shows the WNZ The figure 4.35(c) shows the HAZ. On the serial number 9, FS welding were performed with the tool spindle speed rotation of 1500 rpm with 3^0 tool tilt angle and 30 mm per min. table traverse speed which yields lowest Ultimate Tensile strength of 198 Mpa. with joint efficiency of 48.52% due to the comparatively smaller bead width by the higher tool rotation speed with faster travel speed with wider tool tilt angle weld material consolidation inter related micro chemistry.

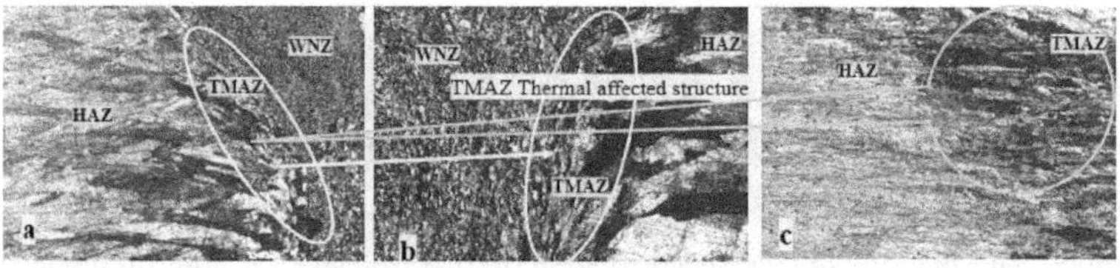

Figure 4.35 S.No.9 Hex Ball tool WNZ, TMAZ, HAZ Thermal affected microstructure

4.5.7 Microstructural characterization of octagonal pin profile pin weldments

Microstructures of different weld zone of octagonal pin profile trial-2 weld joint are presented in the Figure 4.36 (a-e). The Figure 4.36(e) shows the WNZ portion of ultrafine DRX recrystallized grain structure. The WNZ grains were very finer than square and hexagonal pin profile tool same zone structures due to the more nos of flat portions in octagonal pin profile tool. Dynamic recrystallization was caused by local frictional heating and severe plastic strain. During FSW, an increase in the combined effect of frictional heating and plastic deformation took place from TMAZ to nugget (K.V. Jata et.al 2000) The WNZ grain size in trial-2 was comparatively coarser than the trials 6 and 9, which is shown in the subsequent Figure 4.36 (b) and 4.36 (c) respectively. This was confirmed by the same characteristic material flow in experiment 2 and 1 (Venkata Rao et al. (2015), Xu et al (2012)).

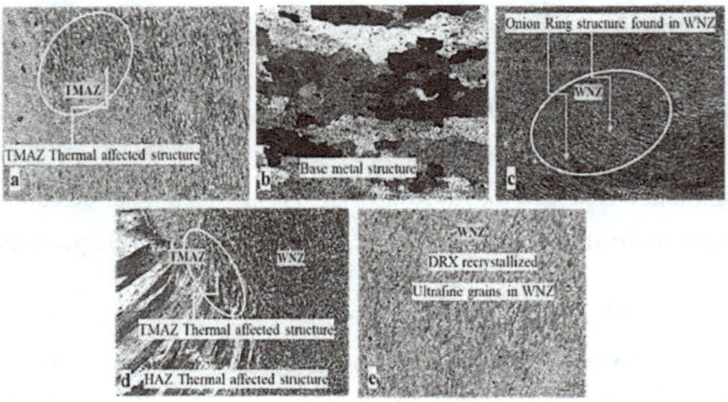

Fig. 4.36. Microstructure of WNZ, TMAZ, HAZ Zones of Octagonal Pin Profile weldment

The subsequent Paragraphs 4.5.7.1, 4.5.7.2 and 4.5.7.3 of microstructural characterization of octagonal pin profile tool deals about the best, moderate and least weld joint result's micro structure morphology in terms of higher tensile strength, higher microhardness and lower corrosion rate from the higher level to lower levels respectively.

4.5.7.1 Microstructural characterization of octagonal pin profile pin-S.no.2 weldments

The figure 4.37 shows the microstructural morphology of S.No.2 Octagonal pin profile various weldment zones. The figure 4.37(a) shows the TMAZ. The figure 4.37(b) shows the WNZ The figure 4.37(c) shows the HAZ. On the serial number 2, FS welding was performed with the tool spindle speed rotation of 1000 rpm with 3^0 tool tilt angle and 25 mm per min. table traverse speed which yields higher Ultimate Tensile strength of 256 Mpa. with joint efficiency of 62.74% due to the higher bead width along with uniformly distributed fine $CuAl_2$ particles joint chemistry with the parent metal. The same were noticed by the previous researcher. (Sahu et al., 2023)

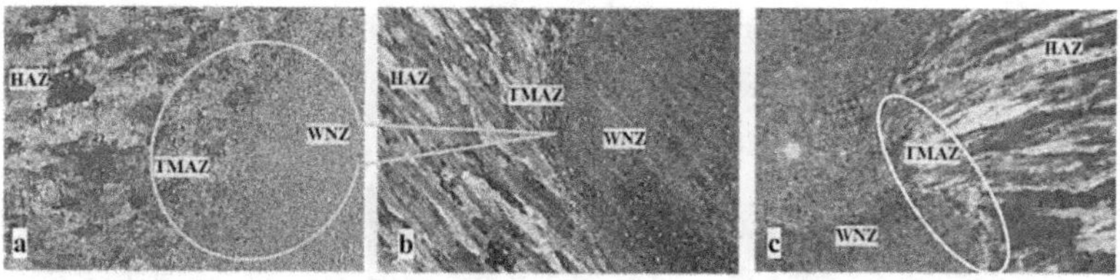

Figure 4.37 SNo.2 OBT Weld specimen WNZ, TMAZ, HAZ Thermal affected microstructure

4.5.7.2 Microstructural characterization of octagonal pin profile pin-S.no.5 weldments

The below figure 4.38 shows the microstructural morphology of S.No.5 Octagonal pin profile various weldment zones. The figure 4.38(a) shows the TMAZ. The figure 4.38(b) shows the WNZ The figure 4.38(c) shows the HAZ. On the serial number 5, FS welding were performed with the tool spindle speed rotation of 1250 rpm with 3.5^0 tool tilt angle and 25 mm per min. table traverse speed which yields moderate Ultimate Tensile strength of 234 Mpa. with joint efficiency of 57.35%. As per the joint no. 5 moderate bead width by the moderate tool rotation speed with tool traverse speed and with higher tool tilt angle along with uniformly distributed fine $CuAl_2$ particles joint chemistry with the parent metal. The same were noticed by the previous researcher. (Sahu et al., 2023)

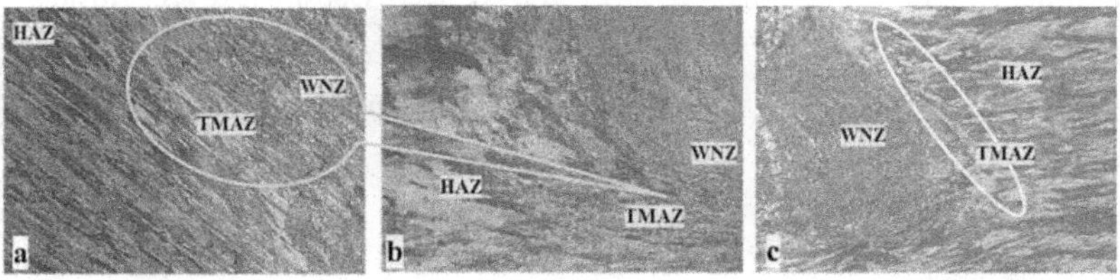

Figure 4.38 SNo.6 OBT Weld specimen WNZ, TMAZ, HAZ Thermal affected microstructure

4.5.7.3 Microstructural characterization of octagonal pin profile pin-S.no.8 weldment

The figure 4.39 shows the microstructural morphology of sno.8 Octagonal pin profile various weldment zones. The figure 4.39(a) shows the TMAZ. The figure 4.39(b) shows the WNZ The figure 4.39(c) shows the HAZ. On the serial number 8, FS welding was performed with the tool spindle speed rotation of 1500 rpm with 2.5^0 tool tilt angle and 25 mm per min. table traverse speed which yields lower Ultimate Tensile strength of 220 Mpa. with joint efficiency of 53.92%. As per the joint no.8, lower bead width by the higher tool rotation speed with tool moderate traverse speed and with moderate tool tilt angle along with uniformly distributed fine $CuAl_2$ particles joint chemistry with the parent metal. The same were noticed by the previous researcher. (Sahu et al., 2023)

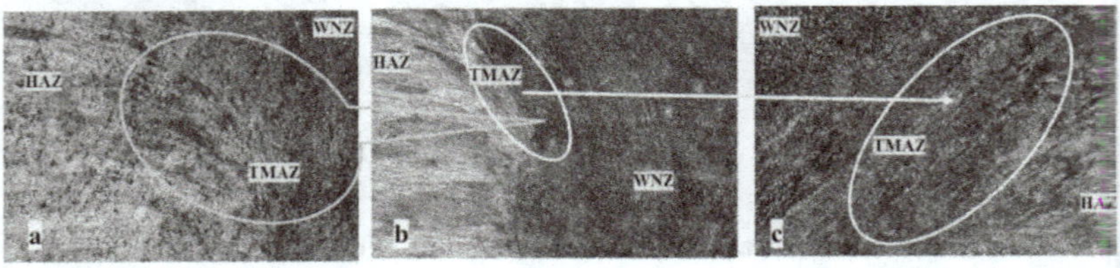

Figure 4.39 SNo.8 OBT Weld specimen WNZ, TMAZ, HAZ Thermal affected microstructure

4.6 Corrosion Testing

In this study different weld zones were exposed to corrosion attack. The corrosion happened due to the potential difference between the anode and cathode, the higher potential difference leads to more Corrosion rate (CR).

4.6.1 Corrosion Rate Calculation for FSW AA2219 weld zone

The corrosion rate for any pin profile AA2219 weld joint, expressed in terms of penetration depth per year, was determined using the standard electrochemical formula:

Corrosion Rate (mm/year) =0.00327×icorr×Eq. Weight Density\text {Corrosion Rate (mm/year)} = 0.00327 \times \frac{i_{\text{Corr}} \times \text {Eq. Weight}}{\text{Density}}Corrosion Rate (mm/year)=0.00327×Densityicorr×Eq. Weight

where:

- i_corr represents the corrosion current density derived from the Tafel plot, measured in $\mu A/cm^2$.
- Eq. Weight denotes the equivalent weight of AA2219, based on the alloy's chemical composition.
- Density is the material density of AA2219, approximately 2.84 g/cm^3.

Under the evaluated conditions, the corrosion current density (i_corr) yielded a calculated corrosion rate of 0.0172 mm/year for AA2219. By substituting the respective values of i_corr, Eq. Weight, and Density into the equation, this rate was obtained, indicating a moderate level of material loss in the corrosive environment.

Table 4.8 Corrosion rate in mm /year of WNZ of square, hexagonal and octagonal pin profile tool weldments Various FSW Tool pin profile

Experiment No.	Spindle speed in RPM	Table Travel Speed in mm/min	Tool Tilt Angle in degrees	Actual Corrosion Rate in mm/ year Sq. pin	Actual Corrosion Rate in mm /year Hex. pin	Actual Corrosion Rate in mm /year Oct. pin
1	1000	20	2.5	0.02752	0.04072	0.03922
2	1000	25	3	0.01699	0.03165	0.03874
3	1000	30	3.5	0.02819	0.04019	0.04219
4	1250	20	3	0.03575	0.03954	0.04954
5	1250	25	3.5	0.03328	0.04018	0.04891
6	1250	30	2.5	0.03184	0.04513	0.04132
7	1500	20	3.5	0.03681	0.04681	0.04681
8	1500	25	2.5	0.03986	0.04982	0.04286
9	1500	30	3	0.04238	0.04313	0.04213

The table 4.8 shows the corrosion rate in mm /year of Weld Nugget Zone of square, hexagonal and octagonal pin profile tool friction stir weldments. The Figure 4.40 shows the anodic polarization curves square pin profile weldments that were obtained from the Corrosion Testing.

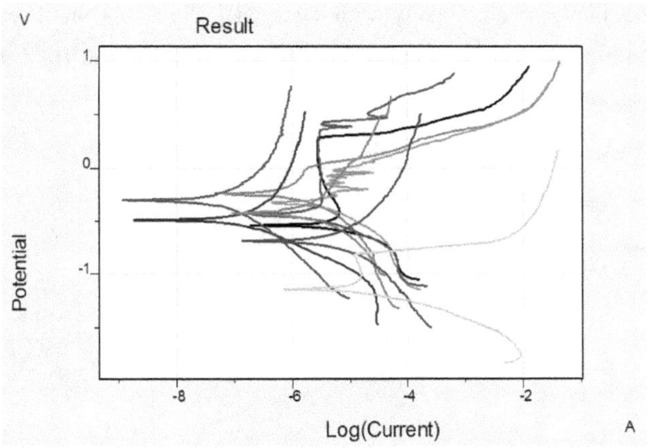

Figure 4.40 Anodic polarization curve (Corrosion plot) of square pin profile weldments

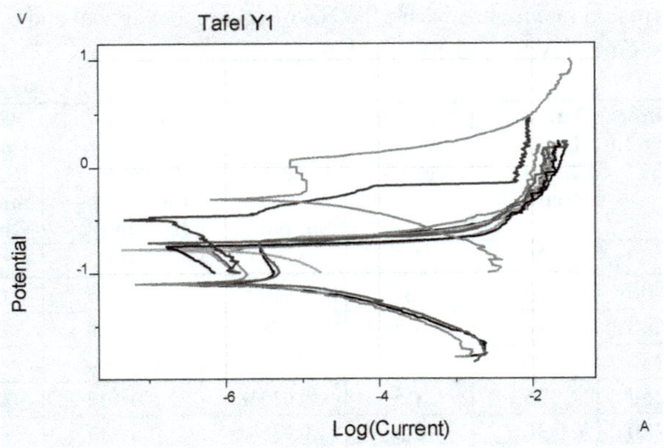

Figure 4.41 Anodic polarization curve (Corrosion plot) for hexagonal pin profile weldments

The Figure 4.41 shows the anodic polarization curves hexagonal pin profile weldments that obtained from the Corrosion Testing. The Figure 4.42 shows the anodic polarization curves for octagonal pin profile weldments were obtained from the Corrosion Testing. From the Potential dynamic corrosion survey for the base metal, the CRA was 0.3616 mm/year, and the HAZ portion was 0.2568 mm/year, but for the weld mid portion the highest corrosion strength value of 0.01699 mm/year was achieved with Square Pin profile tool, subsequently the next high corrosion strength value was achieved with a Hexagonal Pin with strength value 0.03165 mm/year and finally with Octagonal Pin 0.03874mm/year. From the experimental observations, comparing with base metal, TMAZ and HAZ, the corrosion attack was less only in WNZ

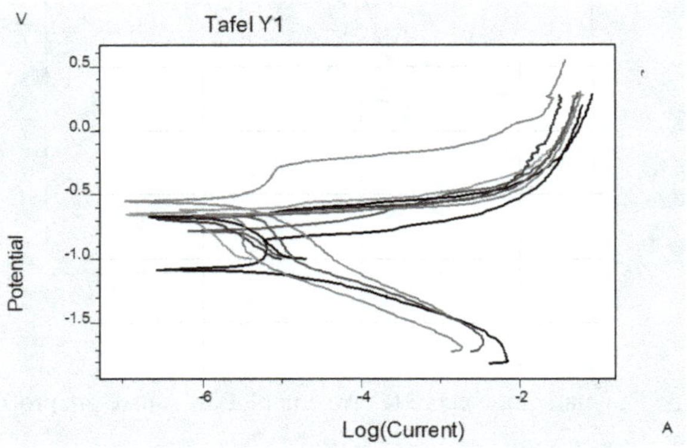

Figure 4.42 Anodic polarization curve (Corrosion plot) for octagonal pin profile weldments

Similar studies on corrosion behaviour in FSW of AA2219-T87 using different tool profiles was conducted by, Rao et al. (2020) and they found that square pin profiles exhibited better corrosion resistance than other shapes due to enhanced grain refinement in the stir zone. Garg et al. (2021) investigated corrosion rates of FSW joints with hexagonal and octagonal pin profiles and reported that hexagonal profiles showed higher corrosion rates due to residual stresses and higher dislocation densities. Kumar et al. (2019) demonstrated that octagonal pins provided a balanced microstructure in the SZ, resulting in intermediate corrosion rates between square and hexagonal profiles. These studies confirm that the corrosion resistance of FSW AA2219-T87 varies with pin profile due to differences in grain structure, dislocation density, and recrystallization within the WNZ, TMAZ, and HAZ.

4.7 Parameter Optimization

A deep network architecture is the SNN architecture. In SNN, the input layer, hidden layers, and output layer make up the model structure as shown in the Fig 4.43. It shows the Input and output Layers of spiking networks.

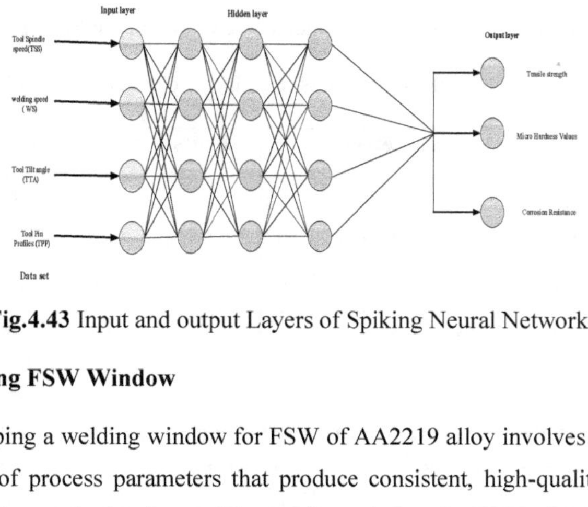

Fig.4.43 Input and output Layers of Spiking Neural Networks

4.7.1 Developing FSW Window

Developing a welding window for FSW of AA2219 alloy involves identifying the optimal range of process parameters that produce consistent, high-quality welds while minimizing defects and other issues. The welding window for FSW of AA2219 typically involves a range of process parameters such as Tool spindle speed-TSS, Welding speed-WS or Tool traverse speed TTS), TTA-Tool tilt Angle and tool geometry. The parameters are selected based on the properties of the material and the desired weld quality. The following list contains the basic processes that make up in the equilibrium optimization algorithm.

1. Randomly establishing the Population

2. Objective Function Assessment

3. Fitness function finding.

4. Up until the method converges to the ideal point, apply genetic operations such as crossover, mutation, and random selection iteratively. It used the goal Function to carry out the SNN optimization process.

5.Dimensions of the population produced at random: 100.

6.Crossover percentage: 0.9. mutation rate: 0.01.

7.Random selection is a selection criterion.

The following relationship is used to determine the intermediate coded values. The rolled AA2219 plates of size 100 mm by 100 mm by 6.2 mm thick were used as the basic raw material, To determine the upper and lower bounds of the process parameters for the 6.2 mm thick aluminium alloy AA2219, trial runs were carried out. The parameters' practical limitations were selected such that the joint should be devoid of obvious flaws. The greatest and bottom limits of the parameter were also coded as +2 and -2, respectively. The table 4.9 shows the various levels of input parameters were used in the FSW process.

Table 4.9 The levels of the Parameters used in the Friction Stir welding process

Symbol	Weld Parameters	Units	Levels		
			1	2	3
X	Tool Spindle Speed (TSS)	rpm	1000	1250	1500
Y	Table Traverse Speed (WS/TTS)	mm/min	20	25	30
Z	Tool tilt Angle (TTA)	degree	2.5	30	3.5

4.7.2 Ultimate Tensile strength, Microhardness and corrosion rate outputs with FSW Input Parameters through SNN Optimization.

This section below provides a detailed discussion about the proposed SNN-EAO experimental method towards the arrival of optimized FSW input parameters in aiming better Ultimate Tensile strength, Microhardness and corrosion rate outputs for the current used FSW tool pin Profiles of square, hexagonal and Octagonal.

Table 4.10 UTS Experimental output and SNN training data output of square Pin profile weldments

S.No.	Tool profile	Spindle speed in RPM	Table Travel Speed in mm/min	Tool Tilt Angle in degrees	Actual Tensile Strength in Mpa	Experimental Value	Difference
1	Square Pin	1000	20	2.5	271	271.6716	0.6716
2	Square Pin	1000	25	3	293	292.5248	0.4751
3	Square Pin	1000	30	3.5	257	256.3462	0.6537
4	Square Pin	1250	20	3	249	248.4695	0.5304
5	Square Pin	1250	25	3.5	276	276.459	0.459
6	Square Pin	1250	30	2.5	249	248.4056	0.5943
7	Square Pin	1500	20	3.5	214	214.5545	0.5545
8	Square Pin	1500	25	2.5	225	225.0249	0.0249
9	Square Pin	1500	30	3	209	208.4798	0.5201

Table 4.11 Average UTS values of square pin weld

Square Pin	Tensile Strength
MAE	3.02E-05
RMSE	3.30E-05
MSE	1.09E-09
R-SQUARE	1
Correlation	1

The proposed model is evaluated with the existing techniques in terms of performance matrices such as RMSE, MAE, MSE, R-Square, and Correlation coefficient. The subsequent tabulations show the Actual Tensile Strength in Mpa of the weld joints results through the proposed SNN model by Square Pin, Hexagonal Pin and Octagonal Pin profile tool, and actual results of ultimate tensile strength of weld specimen by welding with the above specified pin profile weld tool. The table 4.10 shows the ultimate tensile strength of Square Pin profile weldments and The table 4.11 shows the average UTS values of square profile weldments. The following tabulations show the Actual Tensile Strength in Mpa of the weld joints results through the proposed SNN model by hexagonal Pin profile tool, and actual results of ultimate tensile strength of weld specimen by welding with the hexagonal pin profile weld tool. The table 4.12 shows the ultimate tensile strength of hexagonal Pin profile weldments and the table 4.13 shows the average UTS values of hexagonal pin profile weldments.

Table 4.12 UTS Experimental output and SNN training data output of hexagonal Pin profile tool weldments

S.No.	Tool Profile	Spindle speed in RPM	Table Travel Speed In mm/min	Tool Tilt Angle in degrees	Actual Tensile Strength in Mpa	Experimental Value	Difference
1	Hexagonal Pin	1000	20	2.5	210	209.4717	0.52828
2	Hexagonal Pin	1000	25	3	258	258.3428	0.34283
3	Hexagonal Pin	1000	30	3.5	228	227.5034	0.49655
4	Hexagonal Pin	1250	20	3	201	201.4524	0.45241
5	Hexagonal Pin	1250	25	3.5	239	239.1308	0.13081
6	Hexagonal Pin	1250	30	2.5	207	206.5414	0.45856
7	Hexagonal Pin	1500	20	3.5	202	202.3119	0.31191
8	Hexagonal Pin	1500	25	2.5	213	212.8067	0.19321
9	Hexagonal Pin	1500	30	3	198	197.5242	0.47579

Table 4.13 Average UTS values of hexagonal pin weld

Hexagonal Pin	Tensile Strength
MAE	0.000232
RMSE	0.000278
MSE	7.71E-08
R-SQUARE	1
Correlation	1

The following tabulations show the Actual Tensile Strength in Mpa of the weld joints results through the proposed SNN model by Octagonal Pin profile tool, and actual results of ultimate tensile strength of weld specimen by welding with the octagonal pin profile weld tool. The table 4.14 shows the ultimate tensile strength of octagonal Pin profile weldments and the table 4.15 shows the average UTS values of octagonal profile weldments.

Table 4.14 UTS Experimental output and SNN training data output of Octagonal Pin profile tool weldments

S.No	Tool profile	Spindle speed in RPM	Table Travel Speed In mm/ min	Tool Tilt Angle in degrees	Actual Tensile Strength in Mpa	Experimental Value	Difference
1	Octagonal Pin	1000	20	2.5	241	240.86	0.137617
2	Octagonal Pin	1000	25	3	256	255.82	0.175759
3	Octagonal Pin	1000	30	3.5	237	236.56	0.431231
4	Octagonal Pin	1250	20	3	231	230.75	0.244989
5	Octagonal Pin	1250	25	3.5	234	233.34	0.650485
6	Octagonal Pin	1250	30	2.5	241	241.17	0.170467
7	Octagonal Pin	1500	20	3.5	228	228.58	0.588502
8	Octagonal Pin	1500	25	2.5	220	220.17	0.177998
9	Octagonal Pin	1500	30	3	232	232.57	0.569999

Table 4.15 Average UTS values of octagonal pin weld

Octagonal Pin	Tensile Strength
MAE	0.000232
RMSE	0.000278
MSE	7.71E-08
R-SQUARE	1
Correlation	1

The below figure 4.44 illustrates the UTS values in Mpa of square, hexagonal, and Octagonal pin profile weldments with actual and experimental values.

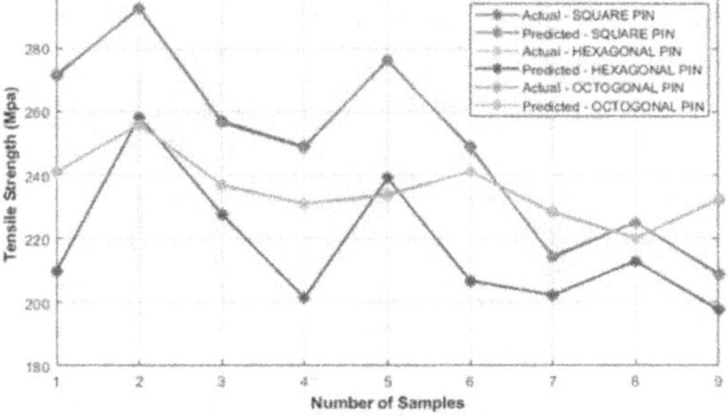

Fig.4.44 Graphical illustration of the UTS values in Mpa of square, hexagonal, and Octagonal pin profile weldments with actual and experimental values

This following section provides a detailed discussion on the proposed SNN-EAO experimental method towards micro hardness, and the proposed model is evaluated with the existing techniques in terms of performance matrices such as RMSE, MAE, MSE, R-Square, and Correlation coefficient. The following tabulations show the actual micro hardness values in Hv of the weld joints results through the proposed model by Square Pin profile tool, and actual results of micro hardness values in Hv of weld specimen by welding with the square pin profile weld tool. The table 4.16 shows the microhardness value in Hv of Square Pin profile weldments and the table 4.17 shows the average microhardness value in Hv of square profile weldments.

Table 4.16: Microhardness values Experimental output and SNN training data output of Square Pin profile tool

S.No.	Tool profile	Spindle speed in RPM	Table Travel Speed In mm/min	Tool Tilt Angle in degrees	Actual Micro Hardness value in Hv	Experimental Value	Difference
1	Square Pin	1000	20	2.5	87.8	87.8256	0.02564
2	Square Pin	1000	25	3	94.6	93.4376	0.1623
3	Square Pin	1000	30	3.5	91.63	91.4479	0.18208
4	Square Pin	1250	20	3	86.27	86.5354	0.26543
5	Square Pin	1250	25	3.5	92.97	92.8751	0.09488
6	Square Pin	1250	30	2.5	94.03	93.7038	0.32612
7	Square Pin	1500	20	3.5	93.83	94.0816	0.25165
8	Square Pin	1500	25	2.5	93.63	93.3301	0.2998
9	Square Pin	1500	30	3	93.6	93.9778	0.3778

Table 4.17: Average Microhardness values of square pin weld

Square Pin	Micro Hardness value
MAE	1.55E-05
RMSE	1.83E-05
MSE	3.35E-10
R-SQUARE	1
Correlation	1

The following tabulations show the Actual microhardness value in Hv of the weld joints results through the proposed SNN model by hexagonal Pin profile tool, and actual results of microhardness value in Hv of weld specimen by welding with the hexagonal pin profile weld tool. The table 4.18 shows the microhardness value in Hv of hexagonal Pin profile weldments and the table 4.19 shows the average microhardness value in Hv of hexagonal profile weldments.

Table 4.18 Microhardness values Experimental output and SNN training data output for Hexagonal Pin profile tool

S.No.	Tool profile	Spindle speed in RPM	Table Travel Speed in mm/min	Tool Tilt Angle in degrees	Actual Micro Hardness value in Hv	Experimental Value	Difference
1	Hexagonal Pin	1000	20	2.5	84.17	84.1309	0.03902
2	Hexagonal Pin	1000	25	3	82.97	82.7374	0.23253
3	Hexagonal Pin	1000	30	3.5	84.33	84.3609	0.03093
4	Hexagonal Pin	1250	20	3	83.43	83.1775	0.25241
5	Hexagonal Pin	1250	25	3.5	83.57	83.781	0.21109
6	Hexagonal Pin	1250	30	2.5	82.7	82.2128	0.48717
7	Hexagonal Pin	1500	20	3.5	84.9	84.6835	0.21643
8	Hexagonal Pin	1500	25	2.5	85.13	85.3747	0.24474
9	Hexagonal Pin	1500	30	3	84.8	85.104	0.30405

Table 4.19 Average Microhardness values of hexagonal pin weld

Hexagonal Pin	Micro Hardness value
MAE	2.51E-04
RMSE	2.88E-04
MSE	8.28E-08
R-SQUARE	1
Correlation	1

The following tabulations show the Actual microhardness value in Hv of the weld joints results through the proposed SNN model by Octagonal Pin profile tool, and actual results of microhardness value in Hv of weld specimen by welding with the above specified pin profile weld tool. The table 4.20 shows the microhardness value in Hv of hexagonal Pin profile weldments and the table 4.21 shows the average microhardness value in Hv of octagonal profile weldments

Table 4.20 Microhardness values Experimental output and SNN training data output for Octagonal Pin profile tool

S.No.	Tool profile	Spindle speed in RPM	Table Travel Speed In mm/min	Tool Tilt Angle in degrees	Actual Micro Hardness value in Hv	Experimental Value	Difference
1	Octagonal Pin	1000	20	2.5	86.43	86.5554	0.1254
2	Octagonal Pin	1000	25	3	87.6	87.1225	0.4774
3	Octagonal Pin	1000	30	3.5	87.03	87.1976	0.1676
4	Octagonal Pin	1250	20	3	90.5	90.2181	0.2818
5	Octagonal Pin	1250	25	3.5	93.14	92.99	0.1499
6	Octagonal Pin	1250	30	2.5	86.08	86.3111	0.2311
7	Octagonal Pin	1500	20	3.5	87.74	87.2576	0.4823
8	Octagonal Pin	1500	25	2.5	89.84	90.1082	0.2682
9	Octagonal Pin	1500	30	3	86.98	86.8972	0.0827

Table 4.21 Average Microhardness values of octagonal pin weld

Octagonal Pin	Micro Hardness value
MAE	2.51E-04
RMSE	2.88E-04
MSE	8.28E-08
R-SQUARE	1
Correlation	1

The figure 4.45 illustrates the microhardness value in Hv of square, hexagonal, and octagonal pin profile weldments with actual and experimental values.

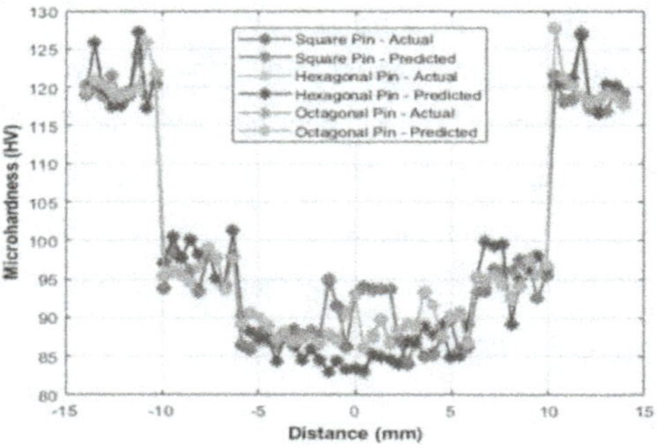

Fig. 4.45 The actual and predicted microhardness values in Hv of various weld zones of square, hexagonal, and octagonal pin profile weldments

Table 4.22 Corrosion Rate value Experimental output and SNN training data output for Square Pin profile tool

S.No.	Tool profile	Spindle speed in RPM	Table Travel Speed In mm/min	Tool Tilt Angle in degrees	Actual Corrosion Rate in mm/year	Experimental Value	Difference
1	Square Pin	1000	20	2.5	0.02752	0.0232	0.0042
2	Square Pin	1000	25	3	0.01699	0.018	0.001
3	Square Pin	1000	30	3.5	0.02819	0.0273	0.0008
4	Square Pin	1250	20	3	0.03575	0.0372	0.0015
5	Square Pin	1250	25	3.5	0.03328	0.0356	0.0023
6	Square Pin	1250	30	2.5	0.03184	0.0329	0.0011
7	Square Pin	1500	20	3.5	0.03681	0.0356	0.0011
8	Square Pin	1500	25	2.5	0.03986	0.0387	0.0011
9	Square Pin	1500	30	3	0.04238	0.0414	0.0009

This following section provides a detailed discussion about the proposed SNN-EAO experimental method towards Corrosion behaviour and the proposed model is evaluated with the existing techniques in terms of performance matrices such as RMSE, MAE, MSE, R-Square, and Correlation coefficient. The following tabulations show the Corrosion Rate values in mm per year of the weld joints results through the proposed SNN model by Square, hexagonal and octagonal Pin profile tool, and actual results of Corrosion Rate values in in mm per year of weld specimen by welding with the above specified pin profile weld tool. The table 4.22 shows the Corrosion Rate value in mm per year of Square Pin profile weldments and the table 4.23 shows the Average Corrosion Rate value in mm per year of square pin profile weldments

Table 4.23 Average Corrosion Rate values of Square Pin weld

Square Pin	Corrosion Rate
MAE	2.57E-05
RMSE	2.98E-05
MSE	8.90E-10
R-SQUARE	1
Correlation	1

The following tabulations show the Actual Corrosion Rate value in mm per year of the weld joints results through the proposed SNN model by Hexagonal Pin profile tool, and actual results of Corrosion Rate value in mm per year of weld specimen by welding with hexagonal pin profile weld tool. The table 4.24 shows the Corrosion Rate value in mm per year of hexagonal Pin profile weldments and the table 4.25 shows the average Corrosion Rate value in mm per year of hexagonal profile weldments

Table 4.24 Corrosion Rate value Experimental output and SNN training data output for Hexagonal Pin profile tool

S.No	Tool profile	Spindle speed in RPM	Table Travel Speed In mm/min	Tool Tilt Angle in degrees	Actual Corrosion Rate in mm/year	Experimental Value	Difference
1	Hexagonal Pin	1000	20	2.5	0.04072	0.04384	0.00312
2	Hexagonal Pin	1000	25	3	0.03165	0.027	0.00464
3	Hexagonal Pin	1000	30	3.5	0.04019	0.03628	0.0039
4	Hexagonal Pin	1250	20	3	0.03954	0.03566	0.00387
5	Hexagonal Pin	1250	25	3.5	0.04018	0.04109	0.00091
6	Hexagonal Pin	1250	30	2.5	0.04513	0.04697	0.00184
7	Hexagonal Pin	1500	20	3.5	0.04681	0.0464	0.0004
8	Hexagonal Pin	1500	25	2.5	0.04982	0.04738	0.00243
9	Hexagonal Pin	1500	30	3	0.04313	0.04656	0.00343

Table 4.25 Average Corrosion Rate values of Hexagonal Pin weld

Hexagonal Pin	Corrosion Rate
MAE	2.10E-04
RMSE	2.40E-04
MSE	5.75E-08
R-SQUARE	1
Correlation	1

The following tabulations show the Actual Corrosion Rate value in mm per year of the weld joints results through the proposed SNN model by Octagonal Pin profile tool, and actual results of Corrosion Rate value in mm per year of weld specimen by welding with octagonal pin profile weld tool. The table 4.26 shows the Corrosion Rate value in mm per year of octagonal Pin profile weldments and the table 4.27 shows the average Corrosion Rate value in mm per year of Octagonal profile weldments

Table 4.26: Corrosion Rate value Experimental output and SNN training data output for Octagonal Pin profile tool

S.No.	Tool profile	Spindle speed in RPM	Table Travel Speed In mm/min	Tool Tilt Angle in degrees	Actual Corrosion Rate in mm/year	Experimental Value	Difference
1	Octagonal Pin	1000	20	2.5	0.03922	0.037908	0.001312
2	Octagonal Pin	1000	25	3	0.03874	0.038312	0.000428
3	Octagonal Pin	1000	30	3.5	0.04219	0.043673	0.001483
4	Octagonal Pin	1250	20	3	0.04954	0.044896	0.004644
5	Octagonal Pin	1250	25	3.5	0.04891	0.051031	0.002121
6	Octagonal Pin	1250	30	2.5	0.04132	0.042914	0.001594
7	Octagonal Pin	1500	20	3.5	0.04681	0.042805	0.004005
8	Octagonal Pin	1500	25	2.5	0.04286	0.042714	0.000146
9	Octagonal Pin	1500	30	3	0.04213	0.044669	0.002539

Table 4.27 Average Corrosion Rate values of Octagonal Pin weld

Octagonal Pin	Corrosion Rate
MAE	2.10E-04
RMSE	2.40E-04
MSE	5.75E-08
R-SQUARE	1
Correlation	1

The previous tables 4.10, 4.12, 4.14, provide actual test results of square, hexagonal and octagonal pin profile weldments Ultimate Tensile strength and further the tables 4.11, 4.13, 4.15. provide the average R-values of Ultimate Tensile strength of the above said pin profiles respectively. The previous tables 4.16, 4.18, 4.20, provide actual test results of of square, hexagonal and octagonal pin profile weldments Microhardness and further the tables 4.17, 4.19, 4.21. provide the average R-values of the Micro hardness of the above said pin profiles respectively. The previous tables 4.22, 4.24, 4.26, provide the Corrosion Rate values and further the tables 4.23, 4.25, 4.27. provides the average R-values of the Corrosion Rate of the square, hexagonal and octagonal pin profile weldments. The experimental outcome result values and the differences between the actual values (AV) and the experimental values (EV) for Ultimate Tensile strength, Micro hardness and corrosion strength of mid weld of different pin experimental samples are compared. The AV represent the desired or target values, while the EV are the measured values obtained from the conducted tests. The differences indicate the deviation between experimental and AV. For each property, the respective tables provide the AV, experimental value, and the difference. Additionally, it shows the average experimental result, average actual result, and the difference between them. From the average R-values tables, it can be observed that there are slight variations between experimental and AV for all three properties. The differences in UTS, microhardness and corrosion strength are within an acceptable range, indicating that the EV are reasonably close to the desired values.

4.8 Confirmation Test

Following the arrival of optimized parameters with the values of the appropriate pin profile tools of Square Pin, Hexagonal Pin and Octagonal pin, the confirmation test welding was carried out and their respective experimental outcome results are tabulated for the comparison of outcome performance amongst the three.

Table: 4.28 Conformation test results of Square Pin

	Spindle speed in RPM	Tool Travel Speed in mm/min	Tool Tilt Angle in degrees	Tensile Strength in Mpa	Microhardness in Hv	Corrosion Rate in mm/year	Average R-Value
Optimal Solution	1132.96	24.6044	3.1231	287.04	93.7064	0.02001	
Feasible Solution	1100	25	3	281.956	94.9737	0.01721	97.3187
Confirmation Test Result	1100	25	3	291.7	97.6	0.0169	
Percentage Error				3.45633	2.7652	1.8221	

The tables 4.28, 4.29 and 4.30 are the confirmation test results of square, hexagonal, and octagonal pin profile tool weldments. By comparing the outcome parameters from the confirmation test results of Ultimate Tensile Strength, Micro Hardness value and the Corrosion Rate, the outperformed tool based on results amongst the three, the best one will be identified for future applications.

The above table 4.28 presents the Conformation test results of Square Pin profile tool welding. The confirmation test yielded with an Ultimate Tensile Strength of 291.70

MPa, a Microhardness value of 97.6 Hv, and a Corrosion Rate of 0.0169 mm/year. The percentage errors for Ultimate Tensile Strength, Microhardness, and Corrosion Rate were 3.45633%, 2.7652%, and 1.8221%, respectively. The average R-value was 97. 3187.

The below table 4.29 presents the Conformation test results of the Hexagonal Pin profile tool welding. The confirmation test yielded with an Ultimate Tensile Strength of 258 MPa, a Microhardness value of 82.97 Hv, and a Corrosion Rate of 0.03465 mm/year. The percentage errors for Ultimate Tensile Strength, Micro Hardness, and Corrosion Rate were 1.24396%, 1.2849%, and 13.375%, respectively. The average R-value was 94. 6986.

Table: 4.29 Conformation test results of Hexagonal Pin

	Spindle speed in RPM	Tool Travel Speed in mm/min	Tool Tilt Angle in degrees	Tensile Strength in Mpa	Micro hardness in Hv	Corrosion Rate in mm/year	Average R-Value
Optimal Solution	1137.38	26.6582	3.1535	249.467	84.0503	0.04389	
Feasible Solution	1100	25	3	254.83	84.05	0.04	94.6986
Confirmation Test Result	1100	25	3	258	82.97	0.03465	
Percentage Error				1.2849	13.375	0.03465	

Table: 4.30 Conformation test results of Octagonal Pin

	Spindle speed in RPM	Tool Travel Speed in mm/min	Tool Tilt Angle in degrees	Tensile Strength in Mpa	Micro hardness in Hv	Corrosion Rate in mm/year	Average R-value
Optimal Solution	1221.82	25.6225	2.6673	248.0947	85.8603	0.046791	
Feasible Solution	1250	25	3	242.39	86.71	0.05	95.6477
Confirmation Test Result	1250	25	3	234	93.14	0.04891	
Percentage Error				3.4613	7.4155	2.18	

The above table 4.30 presents the Conformation test results of the octagonal pin profile tool welding. The confirmation test result yielded with an Ultimate Tensile Strength of 234 MPa, a Micro Hardness value of 93.14HV, a Corrosion Rate of 0.04891 mm/year. The percentage errors for Ultimate Tensile Strength, Micro Hardness, and Corrosion Rate were 3.4613%, 7.41552%, and 2.18%, respectively. The R-value was 95.6477.

4.8.1 Confirmation Testing Outcome of Ultimate Tensile strength using square, hexagonal and octagonal profile tools with optimized parameters.

The figure 4.46 illustrates the results of the confirmation test of Ultimate Tensile Strength values of Square Pin, Hexagonal Pin and Octagonal pin profile tool.

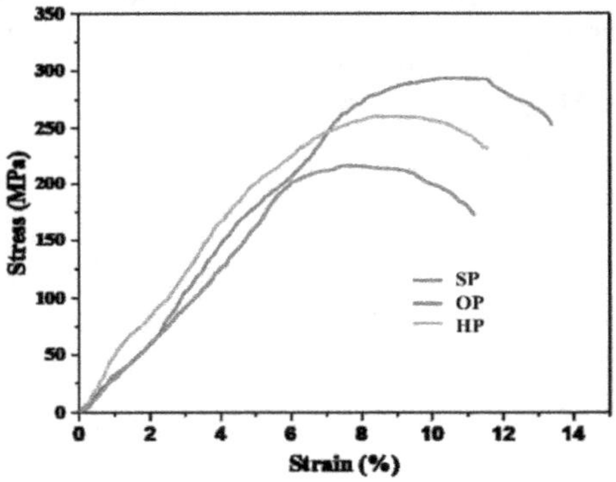

Fig. 4.46 Confirmation testing outcome UTS graphs of the tensile values by the three distinct pin profile tools of square, hexagonal and octagonal ball shoulder tools.

The results of the confirmation test are compared from the tables 4.28, 4.29, and 4.30 of square, hexagonal, and octagonal pins respectively. For the Square profile tool welding, the confirmation test yielded an UTS of 291.70 MPa, a MH of 97.6 Hv, and a CR of 0.0169 mm/year. The percentage errors for TS, MH, and CR were 3.45633%, 2.7652%, and 1.8221%, respectively. The R-value was 97. 3187.

For the Hexagonal Pin profile tool welding, the confirmation test yielded an UTS of 258 MPa, a MH of 82.97 Hv, and a CR of 0.03465 mm/year. The percentage errors for TS, MH, and corrosion strength were 1.24396%, 1.2849%, and 13.375%, respectively. The R-value was 94. 6986.

For the Octagonal Pin profile tool welding, the confirmation test resulted in a TS of 234 MPa, a MH of 93.14Hv, and a CR of 0.04891 mm/year. The percentage errors for TS, MH and CR were 3.4613%, 7.41552%, and 2.18%, respectively. The R-value was 95.6477. These confirmation test findings demonstrate that, based on average R values, all experimental and confirmation test results, the outcomes were within very small margins of error.

4.8.3 Confirmation Testing Outcome of micro structures of square, hexagonal and octagonal pin profile tool weldments

Based on these confirmation test findings, the various output parameters outcome results demonstrate that, the average R values, all experimental and confirmation test results were within very small margins of error. The Square Pin profile tool yielded comparatively the best microhardness, Ultimate Tensile Strength, and corrosion rate results among the three distinct pin profiles.

However, the Hexagonal Pin profile tool produced comparatively lower values of results, with correspondingly lower values of Ultimate Tensile Strength, Micro hardness value, and corrosion strength. The Octagonal Pin profile tool has values in-between the square Pin and Hexagonal Pin tool weldments. The Square Pin profile tool had the best microhardness, Ultimate Tensile Strength, and corrosion rate results out of the previous explained three distinct pin profiles. The figure Fig. 4.47 (b1–b4) shows the Square Pin profile tool confirmation test weld zones microstructure.

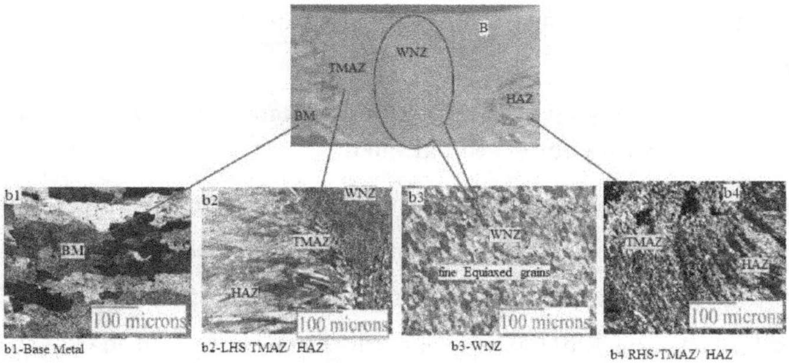

Fig. 4.47 (b1–b4) Square Pin profile tool confirmation test weld zones microstructure

The figure Fig. 4.48 (a1–a4) shows the Hexagonal pin profile tool confirmation test weld zones microstructure.

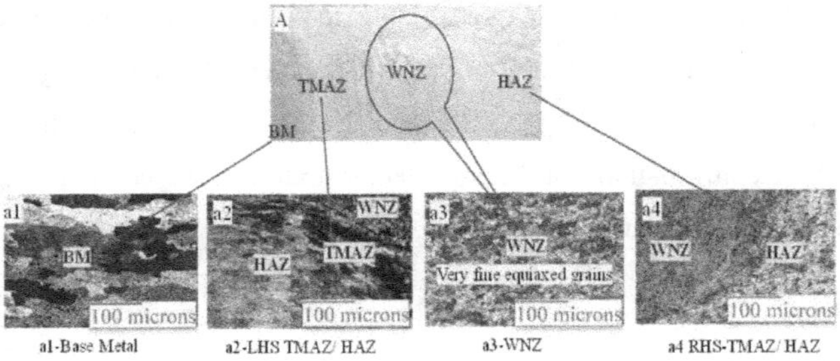

Fig. 4.48 (a1–a4) Hexagonal pin profile tool confirmation test weld zones Microstructure

The figure Fig. 4.49 (c1–c4) Octagonal Pin profile tool confirmation test weld zones microstructures

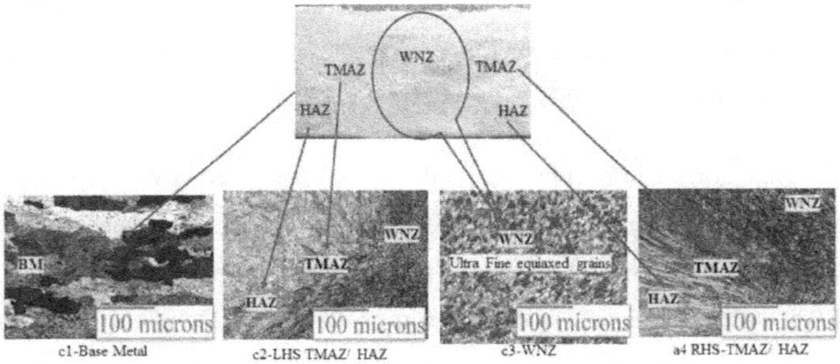

Fig. 4.49 (c1–c4) Octagonal Pin profile tool confirmation test weld zones microstructures

The microstructural arrangements of grains in WNZ, TMAZ and HAZ were almost similar to the previous square, hexagonal and octagonal pin profile weldments S.No.2 of FS Weldments.

4.8.3 Confirmation Testing Outcome of Corrosion Rate using square, hexagonal and octagonal profile tools with optimized parameters.

Based on optimized parameter input values, the confirmation test was conducted with square pin, hexagon Pin, Octagonal Pin profile tool weldments. The CR results of the HP tool profile at a spindle speed of 1100 rpm with a WS of 25 mm /min and TTA of 3^0 the weldment achieved a CR of 0.03465 mm/year. The welding with an OP tool profile at a spindle speed of 1250 rpm, with a WS of 25 mm /min and TTA of 3^0 the weldment achieved a CR of 0.04891 mm/year. The welding with a Square Pin profile tool at a spindle speed of 1100 rpm, with a WS of 25 mm /min and TTA of 3^0 the weldment achieved a CR of 0.0169 mm/year.

From the experimental results, it is evident that for the particular tool pin profiles TSS, WS and TTA had significant impact on microstructure, and the micro structure having significant impact on TS. When the TSS increases, it further leads into High Heating and subsequent softening of weldment and reduction in hardness values. The corrosion immersion test, potentiodynamic test shows the FSW weld joint of WNZ portion having higher corrosion resistance compared to other zones due to the ultra-fine Al matrix grains.

4.8.3.1 Interpretation of Corrosion Rate

The Figure 4.50 shows the Confirmation test Corrosion plot graphical representation of Weld Nugget Zone of Square Pin profile tool. The corrosion rate of 0.0172 mm/year for AA2219 reflects a stable corrosion resistance in the tested conditions.

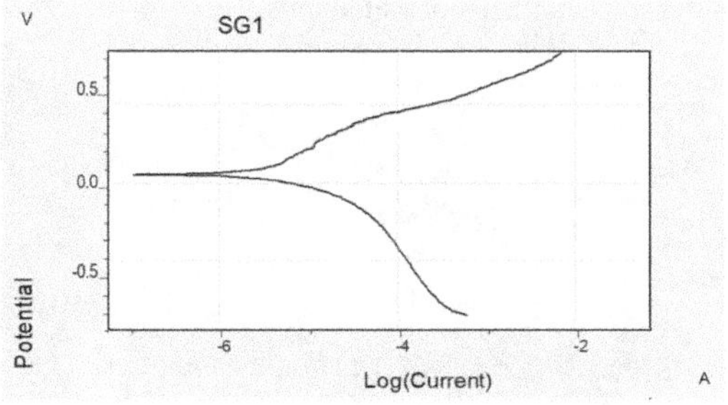

Fig. 4.50 Confirmation test Corrosion plot of Weld Nugget Zone of Square Pin profile tool

Typically, corrosion rates below 0.1 mm/year are deemed acceptable for applications that demand moderate corrosion resistance. This rate suggests that AA2219 is suitable for environments where gradual degradation may occur, yet the material is expected to remain structurally sound over an extended period.

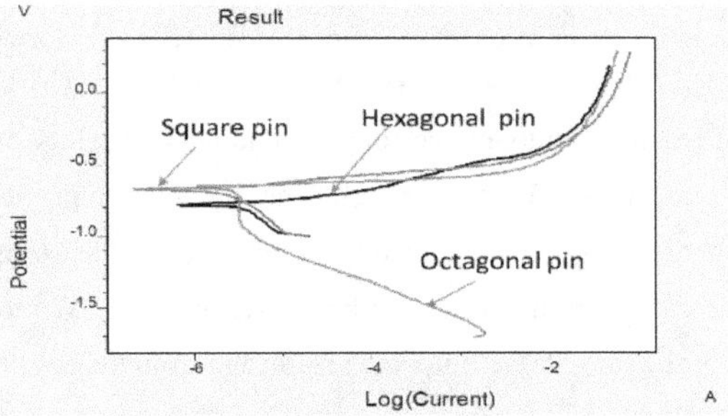

Fig. 4.51. Confirmation test Corrosion plot of Weld Nugget Zone of Square, hexagonal an octagonal pin profile tool

4.8.3.2 Significance of Polarization Behaviour for Optimization of Parameters

The polarization behaviour of AA2219, as depicted in Figure 4.51, infers critical insights not only for corrosion rate estimation but also for optimizing the alloy's performance in corrosive settings. Analysis of the anodic and cathodic slopes allows for the identification of passivation tendencies and susceptibility to localized corrosion forms, such as pitting or crevice corrosion. The linearity of the Tafel regions indicates a predictable corrosion response, which is beneficial for process optimization and alloy selection in applications requiring consistent and manageable corrosion characteristics.

4.9 EDX Microstructure Analysis of FSW AA2219-T87 Square pin weldments

The Energy Dispersive X-ray (EDX) analysis and EBSD analysis were carried out on the best among outcome results yielded tool weldments.

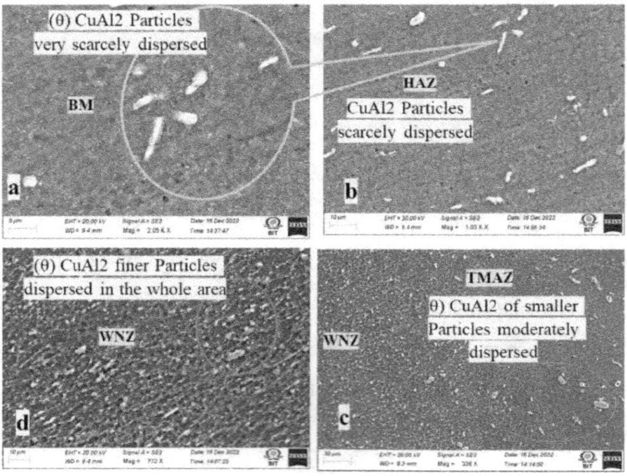

Fig 4.52 Scanning Electron Microscopic morphology (EDAX) show the constituent particles and their size of AA2219-T87 FS weldment of BM, TMAZ, HAZ, WNZ

According to the outcome results of the total 27 numbers of joints from the hexagonal, octagonal and square pin profile weldments, The square pin profile serial number 2 weldments yielded the best outcome results of highest tensile strength, higher micro hardness, and lower corrosion rate results with their respective FSW process parameters of best among the all profile tools weld outcome results and from the confirmation results also confirms the same. Based on comparatively higher tensile strength, higher microhardness, lower corrosion rate EBSD, EDX were performed on the square pin profile welded joint no:2 to analyse the elemental composition of various zones towards identifying grain orientation and grain size formation of different zones for their best outcome results. The elemental analysis helps to assess the material composition changes due to welding heat and deformation, which influences mechanical properties and corrosion resistance. It primarily focuses on identifying the elemental distribution, particularly aluminium and copper. The figure 4.44 shows the SEM microstructural morphology along with EDX analysis show the constituent particles and their size of AA2219-T87 aluminium alloy friction stir weldment by square pin profile, which yields the best outcome results

The 4.52a shows the BM having major α aluminium particles with very scarcely placed bigger (θ) $CuAl_2$ particles, it also got revealed from the below Figure 4.53 EDS analysis of a particular spot analysis of the base metal of AA2219-T87. Here the peak at 1.3 degree shows the rich α aluminium of 91.11% by weight followed by next major copper alloy of 6.92 % by weight , manganese of 0.88% by weight , Titanium of 0.46% by weight, Vanadium of 0.35% by weight and Zirconium of 0.28% by weight.

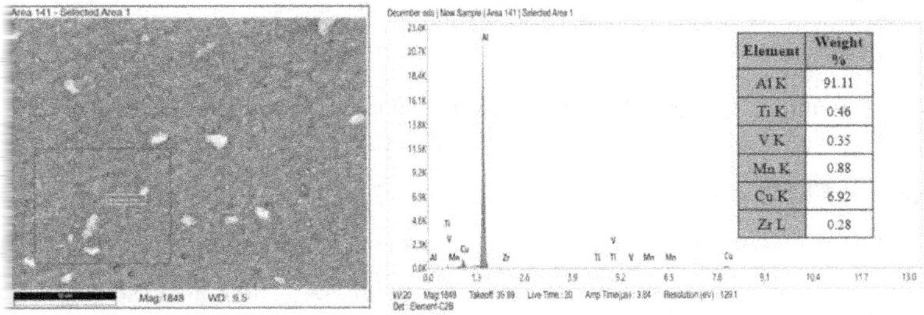

Fig. 4.53 Scanning Electron Microscopic image and EDS analysis of AA2219-T87 alloy (Base Metal)

The Figure 4.53 shows the analysis of a particular portion of the base metal AA2219-T87. It was observed that the AA2219-T87 consisted of a major amount of aluminium followed by copper and smaller quantities of vanadium, zirconium, manganese, and other elements. The recorded values in the EDS analysis were in accordance with the measured elemental composition of AA2219-T87 alloy, as depicted in below Table 4.31.

Table 4.31 Chemical composition of AA2219 alloy Base metal plate

Cu	Mn	Zr	V	Ti	Mg	Si	Zn	Al
6.30	0.38	0.2	0.10	0.06	0.02	0.20	0.10	balance

The figure 4.52(b) shows the HAZ of the same AA2219-T87 square pin profile weldment, having major α aluminium particles with scarcely placed bigger (θ) $cuAl_2$ particles, which is also revealed in the Figure 4.54 EDS spot analysis of the heat affected zone area. The constituent summary of the EDS analysis reveals the more similar availability of α aluminium of 91.75% by weight followed by major copper alloy with 6.70 % by weight then with manganese of 0.66% by weight then further Titanium of 0.39% by weight, Vanadium of 0.43% by weight and Zirconium of 0.07% by weight. This shows that there is no impact of pin and shoulders frictional heating as well as extrusion effects in altering the chemical composition in HAZ. The Figure 4.54 shows the particular spot analysis of the HAZ area of AA2219-T87 by the same square pin profile.

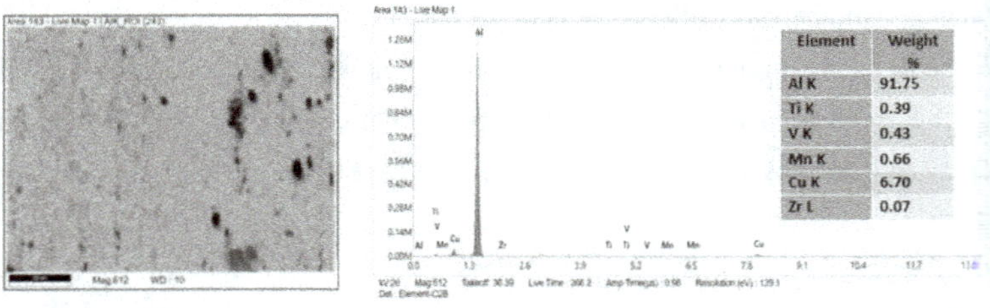

Fig. 4.54 Scanning Electron Microscopic image and EDS spot analysis of AA2219-T87 alloy (HAZ area)

The figure 4.52(c) shows the TMAZ of the same AA2219-T87 square pin profile weldment, having major α aluminium particles with moderately placed comparatively smaller (θ) $CuAl_2$ particles, which is also revealed from the figure 4.46 and it also shows the specific area particular line EDS analysis in the transition zone of TMAZ to WNZ of the thermo mechanically affected area of FSW AA2219-T87 aluminium alloy by the same square pin profile.

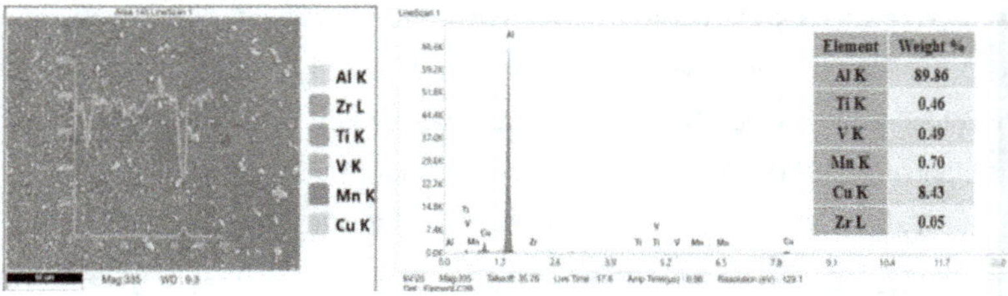

Fig 4.55 Scanning Electron Microscopic image and EDS line graph of AA2219-T87 alloy (TMAZ area)

The Figure 4.55 shows the EDS line analysis of AA2219-T87 through the WNZ to TMAZ transition area. Scanning Electron Microscopic morphology, reveals that there is a minor variation. The constituent summary shows there is a minor variations in the availability of α aluminium of 89.86% by weight followed by major copper alloy with 8.43 % by weight then with manganese of 0.70% by weight then further Titanium of 0.46% by weight, Vanadium of 0.49% by weight and Zirconium of 0.05% by weight. The elemental concentration measurement analysis reveals that there is a thermal influence and frictional heating impacts from the tool shoulder typically alters the constituents and grain size variations in the TMAZ. The below Figure 4.56 shows the EDX spot analysis of AA2219-T87 TMAZ area that further confirms that there is a marginal alteration of constituent elements in TMAZ.

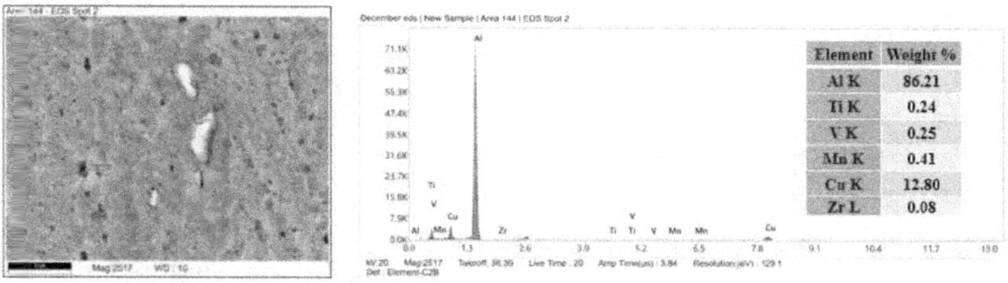

Element	Weight %
Al K	86.21
Ti K	0.24
V K	0.25
Mn K	0.41
Cu K	12.80
Zr L	0.08

Fig 4.56 Scanning Electron Microscopic image and EDS spot analysis of AA2219-T87 alloy (TMAZ area)

The Scanning Electron Microscopic morphology from the figure 4.56 confirms and reveals there is higher constituent elemental variation due to the dynamic spinning effects of spherical ball shoulder in altering the constituent elements with microstructure. The EDS spot analysis of AA2219-T87 TMAZ area and further Scanning Electron Microscopic morphology, reveal there is a higher constituent elemental variations in the availability of α aluminium 86.21% by weight followed by major copper alloy with 12.80 % by weight then with manganese of 0.41% by weight then further Titanium of 0.24% by weight, Vanadium of 0.25% by weight and Zirconium of 0.08% by weight. The elemental concentration measurement analysis reveals there is a thermal influence and frictional heating impacts from the tool shoulder, spherical ball enhancement, which typically alter the constituents and grain size variations in the TMAZ. The basic constituent elements aluminium, copper and other elements were present with marginal change. They were different against as base metal constituent elements weight % as against base metal EDX results.

The figure 4.52(d) shows the WNZ of the same AA2219-T87 square pin profile weldment, having major α aluminium particles with finely dispersed (θ) $CuAl_2$ particles, which also got revealed from the below Figure 4.57. It shows the specific zone area EDS analysis in the WNZ. From the 4.57 EDX elemental graph of the Weld Nugget particular spot area of FSW AA2219-T87 aluminium alloy by the same square pin profile. It is seen that there is a huge variation in the availability of α aluminium of 59.10% by weight followed by major copper alloy with 23.50 % by weight then with manganese of 0.98% by weight then further Titanium of 0.12% by weight, Vanadium of 0.12% by weight and Zirconium of 0.14% by weight. The elemental concentration measurement analysis reveals there is higher dispersion and redissolving of $CuAl_2$ particles by high thermal influence and frictional heating impacts from the tool shoulder with the pin typically altering the constituents and grain size variations in the WNZ.

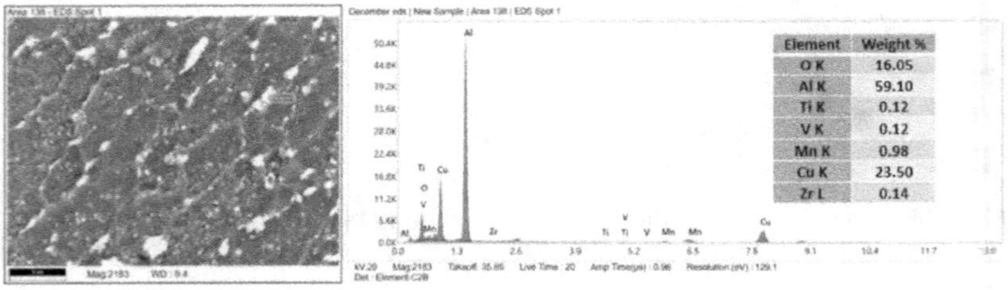

Fig 4.57 Scanning Electron Microscopic image and EDS analysis of AA2219-T87 alloy (WNZ area-Particular Spot)

From the 4.58 EDX elemental graph of the Weld Nugget full zone area of FSW AA2219-T87 aluminium alloy by the same square pin profile. It is observed that there is a variation in the availability of α aluminium of 88.49% by weight followed by major copper alloy with 7.81 % by weight then with manganese of 0.50% by weight then further Titanium of 0.26% by weight, Vanadium of 0.24% by weight and Zirconium of 0.20% by weight. The elemental concentration measurement analysis reveals there is a thermal influence and frictional heating impacts from the tool shoulder with the pin typically altering the constituents and grain size variations in the WNZ.

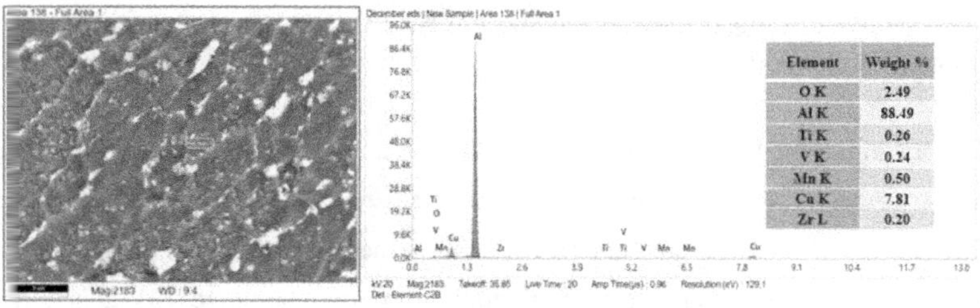

Fig 4.58 Scanning Electron Microscopic image and EDS analysis of AA2219-T87 alloy (WNZ area-Full Zone)

In the weld nugget zone, all the θ'', θ', and some of the θ phase (Al_2Cu) redissolved. Abnormal θ particles were observed in the WNZ, including agglomerated θ with sizes around 100 to 1000 μm and a ring-shaped distribution of normal size θ particles. The formation of abnormal θ particles was resulted from metal plastic flow during welding and the high content of Cu in AA2219. Such redissolution of precipitates caused more Cu in the form of SSS. Compared with θ" and θ' phases, SSS is less effective at hindering dislocation motion, [J.R. Davis et.al 1996 and J.M. Papazian et.al.] causing the low hardness of the WNZ. The form of Cu could be changed with the thermal variation during an FSW process, which caused the different hardness profiles of different regions, causing the low hardness of the WNZ, about 90 HV $_{0.1}$. However, the WNZ went through dynamic recrystallization during welding, which refined the grains and provided some strengthening. (J kang et.al.2016)

Energy Dispersive X-ray Analysis (EDX) of the friction stir welded AA2219-T87 alloy typically examines the distribution and concentration of elements such as aluminium and copper in the weld zone. EDX results often indicate a reduction in copper content within the weld nugget due to material flow and thermal cycles, affecting the distribution of strengthening phases like $CuAl_2$. EDX confirms the changes in alloy composition across the weld zones, helping to explain variations in hardness, corrosion resistance, and mechanical performance in the welded joint (Sahu et al., 2023). Results from such studies often show a reduction in copper content in the weld nugget and redistribution of the strengthening phase particles. These elemental variations directly impact the joint's performance under stress and environmental condition. High copper concentrations at grain boundaries are noted in the nugget, which aids in assessing changes in mechanical and corrosion properties (Khodir et al., 2018).

4.10 EBSD-Metallurgical Characterization of AA2219-T87 Square pin profile Tool friction stir weldments

After Performing the EBSD mapping across the different zones of the friction stir welded samples, namely the base material, heat-affected zone (HAZ), thermo-mechanically affected zone (TMAZ), and stir zone (SZ)). EBSD generates orientation maps that provide information about the crystallographic orientation and grain structure. EBSD patterns can also reveal the grain-boundary and sub-grain-boundary features with high contrast at a sufficient level of magnification.

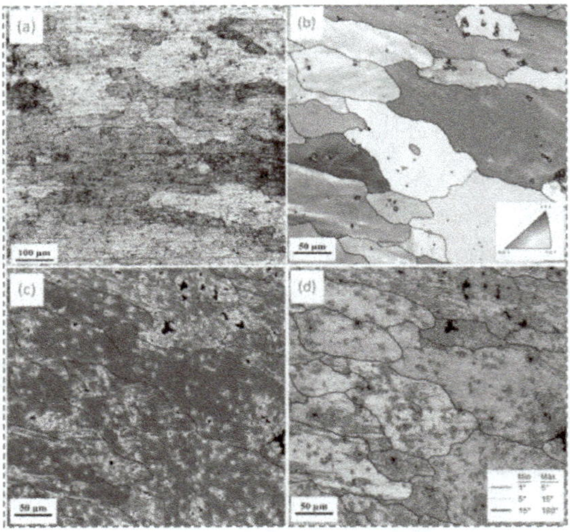

Fig. 4.59 Microstructure of AA2219-T87 BM (a) Optical Microscopic Image (b) EBSD IPF Coloured Map (c) Kernal Average Misorientation Chart and (d) Grain Boundary Map (LAGB's: Less than 15° and HAGB's: More than 15°)

The figure image 4.59 shows the microstructural characteristics of AA2219-T87 base material observed through an optical microscope (Fig. 4.59(a)) and EBSD analysis (Figure 4.59 b–d). From the surface microstructure recorded in the BM Fig. 4.59(a), a flake-like grain structure was observed, the grain size was recorded to be larger, and more intermetallic compounds were dispersed inside the grains. The large grain size of the AA2219-T87 BM was confirmed through the IPF map derived from the EBSD study as depicted in Fig. 4.59(b). Elongated grain structure was noticed in the AA2219-T87 BM during the FSW process and appeared as flake-like grains in the BM of AA2219-T87 alloy.

Moreover, Fig. 4.59(c) shows the Kernal average misorientation (KAM) Map extracted from the EBSD study. Qualitatively, the KAM Map indicates a lower KAM value signifying a lower geometrically necessary dislocation (GND) density. The lower the GND density indicating a lower restriction was developed in the grain boundary. To check the grain boundary distribution, and the amount of lower angle grain boundaries (LAGB's) and Higher Angle Grain Boundaries (HAGB's), a grain boundary distribution map was taken out from the EBSD analysis and presented in Fig. 4.50(d).

In general, a misorientation angle of greater than 15° (recorded in blue colour in Fig. 4.59(d) indicates HAGBs, which corresponds to fine grains, while less than 15° (recorded in red and green colour in Fig. 4.59d) indicates LAGBs. From the grain boundary distribution map (Fig. 4.59d), it was confirmed that the BM of FSW processed AA2219-T87 consists more amount of LAGB's and a comparatively lesser amount of HAGB's, which indicate a lower restriction produced in the grain boundaries. A similar observation was made by the previous researcher (Yao, Xiuquan et.al. 2017).

4.10.1 EBSD-Metallurgical Characterization of AA2219-T87 Weld Nugget Zone (WNZ)

Fig. 4.60 Microstructure of AA2219-T87 WNZ (a) EBSD IPF Coloured Map (b) Kernal Average Misorientation Chart and (c) Grain Boundary Map

Figure 4.60 displays the EBSD microstructural characteristics of WNZ on the FSW-processed AA2219-T87 alloy. Figure 4.60(a) depicts the EBSD IPF Coloured Map and different colours in the IPF map represent different grain orientations and it was noted that the grain size of the WNZ is much smaller when compared to the BM as shown in Figure 4.59(b). High plastic deformation and elevated temperatures during the FSW process triggered dynamic recovery and recrystallization, resulting in the formation of new equiaxed grains in the WNZ with severe plastic flow (Yao, Xiuquan et.al (2023)). The WNZ is affected by the pin and the shoulder and it is observed that the novel spherical ball shoulder makes the grain structure even finer and the size of the grains in the WNZ of the AA2219-T87 was observed.

During the FSW process, the combination of high plastic deformation and elevated temperature results in the development of a larger quantity of recrystallized fine and homogeneous equiaxed grains in the WNZ. Similar outcomes were observed by the previous researchers (Rambabu et.al (2015), Anderson-Wedge et.al (2021)). Moreover, a significantly lower average grain size of 8.86 μm was observed in the WNZ, as depicted in Figure 4.61(b) when compared to the average grain size of BM (52.27μm) as displayed in Figure 4.61(a).

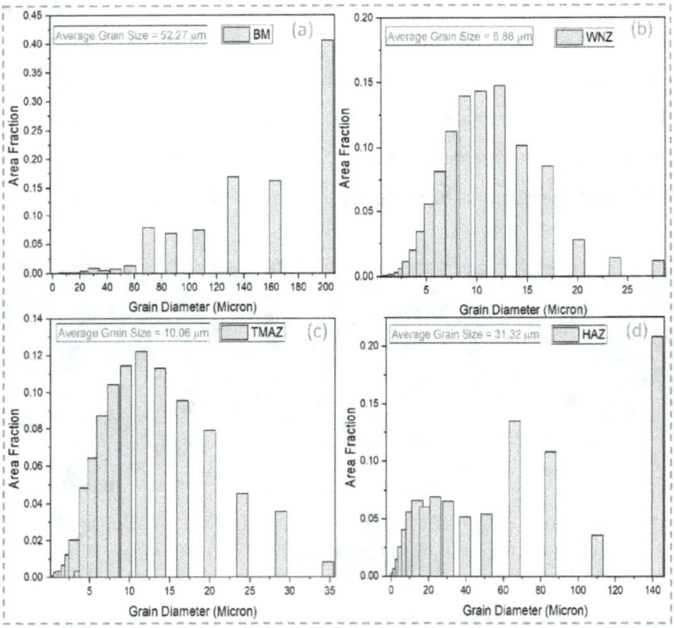

Fig. 4.61 EBSD grain size distribution chart of AA2219-T87 (a) BM (b) WNZ (c) TMAZ and (d) HAZ

4.10.2 EBSD-Metallurgical Characterization of AA2219-T87 Thermo mechanically affected Zone (TMAZ)

Figure 4.62(a), depicts the EBSD IPF coloured Map, figure 4.62(b) depicts the grain boundary distribution map was taken out from the EBSD analysis and figure 4.62(c) shows the Kernal average misorientation (KAM) Map extracted from the EBSD study for the FSW processed AA2219-T87 in the TMAZ for the samples processed with spherical ball shoulder square pin profile tool.

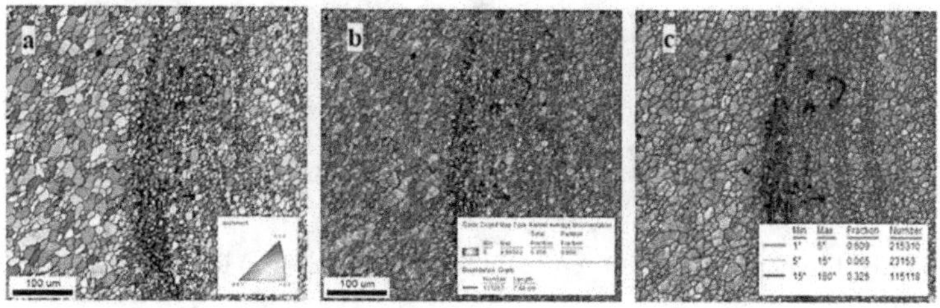

Fig. 4.62 EBSD Microstructure of AA2219-T87 TMAZ grains (a) EBSD IPF Coloured Map (b) Kernal Average Misorientation Chart and (c) Grain Boundary Map

The Fig. 4.62(a) shows the IPF microstructure evolution in TMAZ (TMAZ on AS). It is illustrated that the parent grains tend to be reoriented along the flow pattern around the pin due to shear deformation induced by the rotating tool, which is the typical feature of TMAZ. Newly-formed equiaxed grains are obviously observed, indicating that partial dynamic recrystallization (DRX) occurs. A tendency can be seen from the outside to inside of the TMAZ (i.e. from right to left in Fig. 4.52a), more equiaxed grains are formed. with approaching closer to weld centre, higher temperature and larger locational deformation are achieved, consequently the extent of DRX increases. In the interiors of initial grains, plenty of sub grains and crystallites with partial LABs and HABs can be observed in Fig. 4.5 a and 4.53c. According to Gourdet's study, the crystallite is the intermediate structure between subgrain and grain structure, which is formed during CDRX through gradual LAB to-HAB transformation (Yigezu et.al. (2014)). On pre-existing grain boundaries. Comparatively, the grains of the TMAZ were stretched, the average grain size of 10.06 μm was observed in the TMAZ which may be due to the plastic flow in the WNZ and the thermal cycle of the FSW process.

4.10.3 EBSD-Metallurgical Characterization of AA2219-T87 Heat affected Zone (HAZ)

The HAZ had almost an identical microstructure, when compared to the BM and the size of the grains was notably coarser in the HAZ when compared to the TMAZ as a result of the thermal cycle during the FSW process.

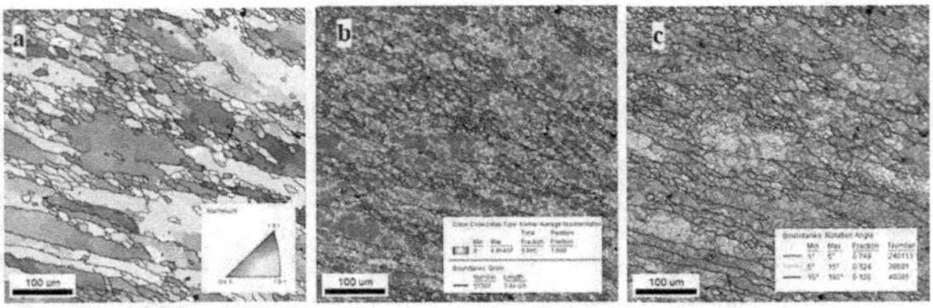

Fig. 4.63 EBSD Microstructure of AA2219-T87 HAZ grains (a) EBSD IPF Coloured Map (b) Kernal Average Misorientation Chart and (c) Grain Boundary Map

Figure 4.63(a) depicts the EBSD IPF coloured Map, figure 4.63(b) depicts the grain boundary distribution map taken out from the EBSD analysis and figure 4.63(c) depicts the Kernal average misorientation (KAM) Map extracted from the EBSD study for the FSW samples processed AA2219-T87 in the HAZ with spherical ball shoulder square pin profile tool. From the EBSD coloured map, it was noted that the grain size was much larger than the TMAZ and it may be ascribed to the welding thermal cycle. The average grain size of HAZ for the FSW processed AA2219-T87 was noted as 31.32 μm as depicted in Figure 4.61(d). In connection with that, moderate KAM values were observed from the KAM chart (Figure 4.63b) This might be attributed to the coarser grains and the development of non-uniformly dispersed precipitates during the welding process. Moreover, a higher amount of subgrain boundaries (1 to 5°) was noticed in the grain boundary chart in the HAZ, which is indicated in the red colour in Figure 4.63(c). The outcomes of the grain boundary chart in the HAZ indicates a poor orientation of the grain boundary when compared to the TMAZ. The HAZ had almost an identical microstructure, when compared to the BM and the size of the grains was notably coarser in the HAZ when compared to the TMAZ as a result of the comparatively lower thermal impact during the FSW process.

4.11 Comparison of square pin plain shoulder tool FS weldments with square pin Spherical ball shoulder tool FS weldments.

Based on the previous FSW processed AA2219-T87 plates with square, hexagonal and octagonal pin profile tool with spherical ball shoulder tool weldments, The outcome microhardness, ultimate tensile strength and corrosion rate results of the above mentioned three different tool pin profile, The square pin profile with spherical ball shoulder tool outperformed with the other pin profile tool weldments. By which the square pin spherical ball shoulder tool was selected for further comparative study with the Plain shoulder square pin profile tool. As per comparative study, welding was done with the SNN optimized feasible parameter values and further output microstructure, microhardness and tensile strength were analysed and the outcome results were presented in the below paragraphs.

4.11.1 Metallurgical Characterization of FSW AA2219-T87 weldments by square pin profile tool plain shoulder and with Spherical Ball Shoulder

The macrostructure (Figs. 4.64 a and b) and the microstructure of the FSW processed AA2219-T87 without and with spherical ball shoulder on the WNZ, TMAZ, and HAZ was revealed in Figure 4.64 which indicates a refinement in the grain structure in the WNZ as shown in Figure 4.64(1) when compared to the Base Metal of AA2219-T87 shown in Fig. 4.59(a). Comparatively smaller equiaxed grain structure was noted in the WNZ of the FSW processed AA2219-T87 without spherical ball shoulder and almost an identical microstructure with an equiaxed grain structure was observed in the FSW processed AA2219-T87 with spherical ball shoulder.

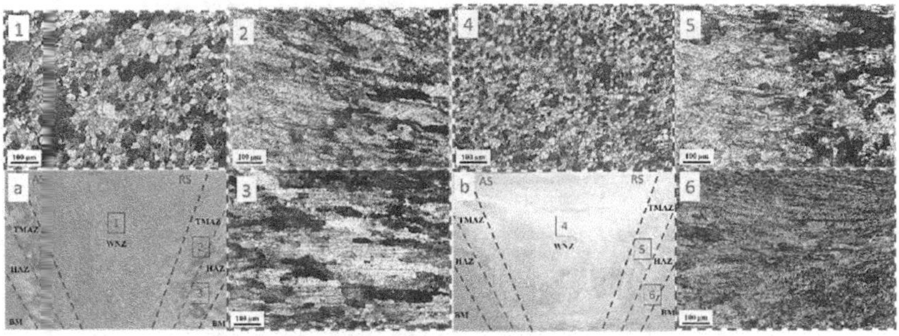

Fig. 4.64 Macrostructure and Microstructure of FSW processed AA2219-T87 without (1-3) and with (4-6) spherical ball shoulder; (1&4) WNZ, (2&5) TMAZ (3&6) HAZ

During the FSW process, the combination of high plastic deformation and elevated temperature results in the development of a larger quantity of recrystallized fine and homogeneous equiaxed grains in the WNZ. Similar outcomes were observed by the previous researchers (Wang et al.2018, Rambabu et al.2015). Moreover, a significantly lower average grain size of 8.86 µm was observed in the WNZ, as depicted in Figure 4.6 (b) when compared to the average grain size of BM (52.27 µm) as displayed in Figure 4.6 (a). Comparatively, the grains of the TMAZ were stretched, which may be due to the plastic flow in the WNZ and the thermal cycle of the FSW process. The HAZ had almost an identical microstructure, when compared to the BM and the size of the grains was notably coarser in the HAZ when compared to the TMAZ as a result of the thermal cycle during the FSW process.

4.11.2 Metallurgical Characterization of AA2219-T87 FS Weld Nugget Zone (WNZ) by square pin profile plain shoulder and special spherical ball shoulder tools

Figure 4.65 displays the microstructural characteristics of WNZ on the FSW processed by square pin plain as well as spherical ball shoulder tool on AA2219-T87 alloy. Figure 4.65(a) shows the EBSD IPF Coloured Map and different colours in the IPF map represents different grain orientations and it was noted that the grains size of the WNZ is much smaller when compared to the BM as shown in the Figure 4.59(b).

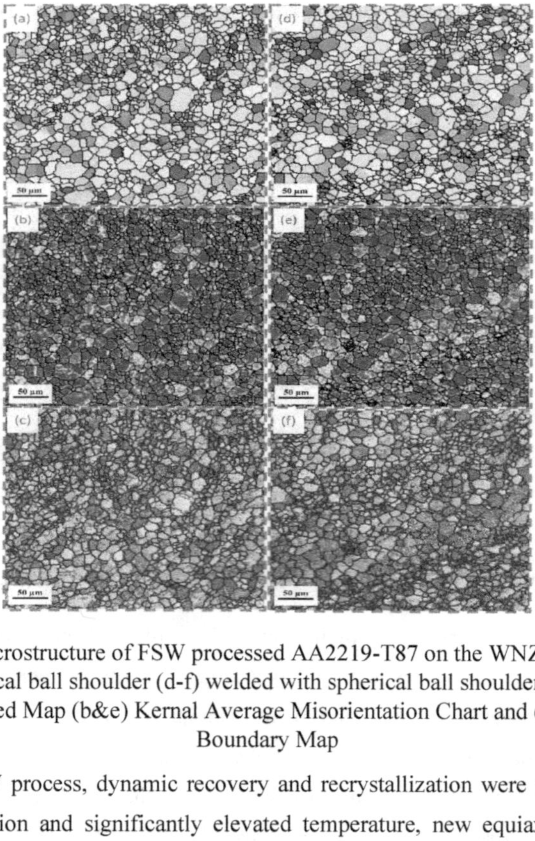

Fig. 4.65 Microstructure of FSW processed AA2219-T87 on the WNZ: (a-c) welded without spherical ball shoulder (d-f) welded with spherical ball shoulder; (a&d) EBSD IPF Coloured Map (b&e) Kernal Average Misorientation Chart and (c&f) Grain Boundary Map

During the FSW process, dynamic recovery and recrystallization were induced by high plastic deformation and significantly elevated temperature, new equiaxed grains were formed in the WNZ with severe plastic flow (Elangovan et al. 2008). The WNZ is affected by the pin and the shoulder and it is observed that the novel spherical ball shoulder makes the grain structure even finer and the size of the grains in the WNZ of the AA2219-T87 was observed to be much smaller than the sample processed without spherical ball. Further, to confirm the misorientation and grain boundary distribution, the KAM chart and grain boundary map was derived from the EBSD study and presented in the Figure 4.65(b-c) and Figure 4.65(e-f) for FS Weld AA2219-T87 without and with spherical ball shoulder respectively. Moderate KAM values were observed from the KAM chart, which indicates a slight increase in the GND density. This increase in the GND density indicates an increase in the restrictions developed in the grains, which is further confirmed through grain boundary distribution chart. From the Figure 4.65(c and f), it is confirmed that more HAGB's were observed for the WNZ when compared to the BM and comparatively the amount of HAGB's were increased in the FS Weld AA2219-T87 sample processed with specially designed spherical ball shoulder. The modification in the shoulder design with the spherical ball improves the grain refinement through the outcome of friction stir processing after the rotation of square pin in the WNZ. The combined effect of FSW and FSP makes the recrystallization, refinement in the grain and reprecipitation of the secondary phases in the WNZ.

4.11.3 Metallurgical Characterization of AA2219-T87 TMAZ and HAZ by square pin profile plain shoulder and special spherical ball shoulder tools

Figure 4.66(a) and 4.66(d) depicts the EBSD IPF coloured Map for the FSW processed AA2219-T87 on the TMAZ welded with plain shoulder and with spherical ball shoulder square pin profile tool.

From the EBSD IPF Coloured Map, it was observed that the FSW process alter the grains more significantly in the TMAZ as well as in the HAZ (Figure 4.67). The mean grain size for the TMAZ was recorded as 10.06 μm and comparatively larger grain size was observed in the HAZ with the measure value of 31.32 μm as shown in the Figure 4.61(c-d). Similarly moderate KAM values were observed from the KAM chart recorded in the TMAZ of AA2219-T87 and it indicates a moderate GND density. From the Figure 4.66(e), it was observed that, more contrasting KAM values in the samples processed with spherical ball shoulder. Qualitatively, the amount the KAM values were observed to be increased than the samples processed without spherical ball shoulder and more importantly, the GND's were concentrated on the grain boundary, which in turn increases the restrictions produced by the grain boundaries.

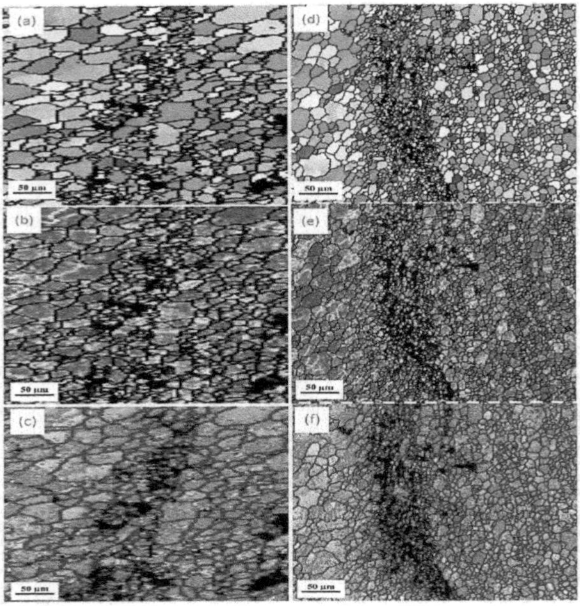

Fig. 4.66 Microstructure of FSW processed AA2219-T87 on the TMAZ: (a-c) welded without spherical ball shoulder (d-f) welded with spherical ball shoulder; (a&d) EBSD IPF Coloured Map (b&e) Kernal Average Misorientation Chart and (c&f) Grain Boundary Map

Besides, the amount of HAGB's was observed to be maximum for the samples processed with spherical ball shoulder when compared to the sample processed without spherical ball shoulder.

Figure 4.67(a) and (d) depicts the EBSD IPF coloured Map for the FSW processed AA2219-T87 on the HAZ for the samples processed without and with spherical ball shoulder respectively. From the EBSD coloured map, it was noted that the grain size was much larger than the TMAZ and it may be ascribed to the welding thermal cycle. The average grain size of HAZ for the FSW processed AA2219-T87 was noted as 31.32 μm as depicted in Figure 4.61(d). In connection with that, moderate KAM values were observed from the KAM chart from Figure4.67(b) and (e), and the amount of LAGB's was observed to be maximum for both the FSW processed AA2219-T87 samples without and with spherical ball shoulder as depicted in the Figure 4.67(c) and (f). This might be attributed to the coarser grains and the development of non-uniformly dispersed precipitates during the welding process.

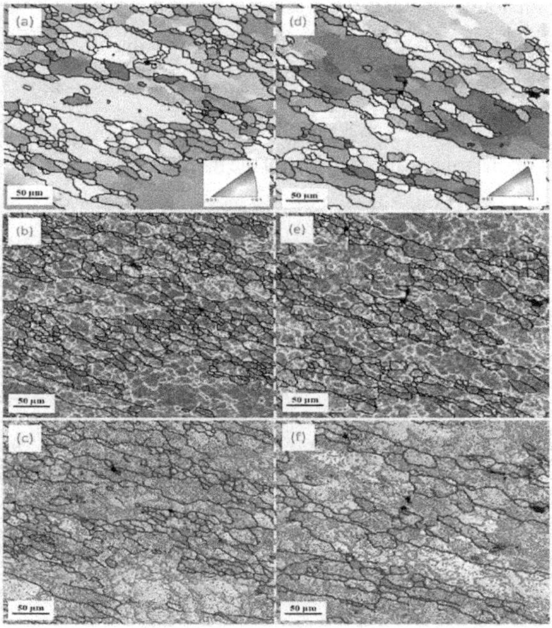

Fig. 4.67 Microstructure of FSW processed AA2219-T87 on the HAZ: (a-c) welded without spherical ball shoulder (d-f) welded with spherical ball shoulder; (a & d) EBSD IPF Coloured Map (b & e) Kernal Average Misorientation Chart and (c & f) Grain Boundary Map

Moreover, a higher amount of subgrain boundaries (1 to 5°) was noticed in the grain boundary chart in the HAZ, which is indicated in the red colour in Figure 4.67 (c) and (f). The outcomes of the grain boundary chart in the HAZ indicates a poor orientation of the grain boundary when compared to the TMAZ, where the amount of subgrain boundaries was found to be minimal in TMAZ, especially for the samples processed with spherical ball shoulder. This clearly indicates that the modification in the tool pin shoulder design improves the grain boundary orientation in the WNZ and TMAZ for the AA2219-T87 samples.

4.11.4 Mechanical Performance of FS Welded AA2219-T87 samples by square pin profile plain shoulder and special spherical ball shoulder tools

The Figure 4.68 depicts the hardness value measured across the weld at different zones of the AA2219-T87 samples processed with plain shoulder and with spherical steel ball shoulder. The hardness value in WNZ of the specimens produced with spherical ball shoulder was comparatively higher than the hardness values of the specimens produced without spherical ball shoulder. Moreover, the hardness value of the TMA and HA zones of the AA2219-T87 alloy processed with spherical shoulder was increased than the sample processed without spherical ball shoulder. The hardness of the FSW sample was greatly influenced by the grain structure and in general the hardness values are directly proportional to the grain size (Heramo et al 2023). As discussed earlier, the grain size of the WNZ for the AA2219-T87 samples processed with spherical ball shoulder was comparative smaller than the samples processed without spherical ball shoulder. The size of the grains was significantly affected by the application of spherical ball shoulder on the WNZ, TMAZ, HAZ and it can be evidenced through the Figures 4.65 - 4.67. Comparatively, the value of hardness in the TMA and HA zones was reduced than the WNZ and it may be ascribed to the softening of the weld zone due to welding thermal cycle during thermo-mechanical processing.

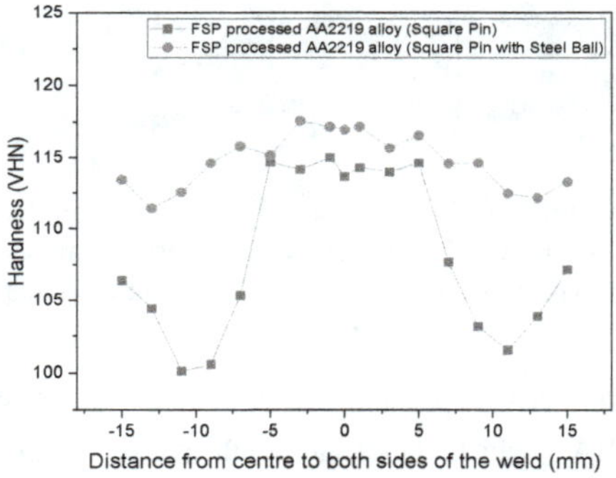

Fig. 4.68 Vickers Hardness values measurement at different distances from the centre of the weld by square pin plain shoulder and with spherical ball shoulder tool

In addition to the grain size, the heat generated during the FSW process, softens the material, dissolving of strengthening precipitates and reprecipitation of the secondary phases plays a vital role determining the hardness values on different zones of the weld material. The possible reason for the reduction in the hardness on the TMA and HA zones may be attributed the coarse elongated grains and reprecipitation of the secondary phases, which in turn reduces the GND densities, which affects the resistance to the load applied.

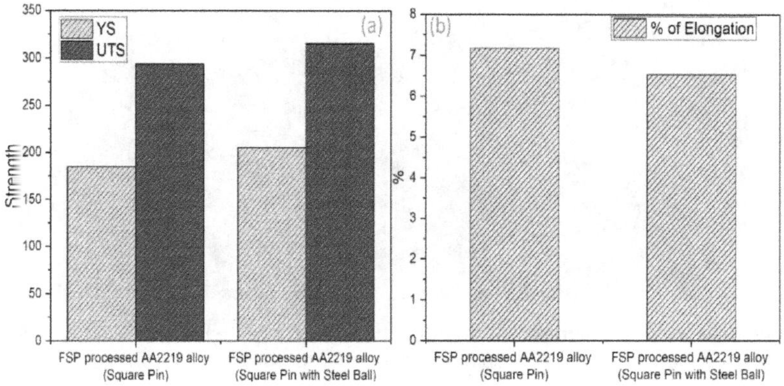

Fig. 4.69 Tensile properties of AA2219-T87 samples processed by square pin tool plain shoulder and with spherical ball shoulder (a) YS and UTS (b) % of Elongation.

Figure 4.69 depicts the tensile behaviour of the AA2219-T87 samples processed without and with spherical ball shoulder. The yield and ultimate tensile strength (YS and UTS) of AA2219-T87 specimens processed with spherical ball shoulder were comparatively higher than the samples produced without spherical ball shoulder as depicted in Figure 4.69. The YS and UTS of the AA2219-T87 specimens processed with spherical ball shoulder were improved by 11.17% and 7.39% than the samples processed without spherical ball shoulder. In contrast to the YS and UTS, the percentage of elongation was slightly reduced than the sample processed with the spherical ball. Figure 4.70 displays the SEM fractography of both the specimens prepared and from the SEM fractography, clear evidence of dimples was noticed in the FSW-processed AA2219-T87 samples processed without and with spherical ball shoulder which confirms that the fracture mechanism was ductile (Wang et al. 2018). From the SEM fractography, it was observed that a greater number of deep and wider dimples were noticed in the FSW processed AA2219-T87 samples processed without spherical ball shoulder, and on the other hand fine dimples were noticed in the samples processed with spherical ball shoulder.

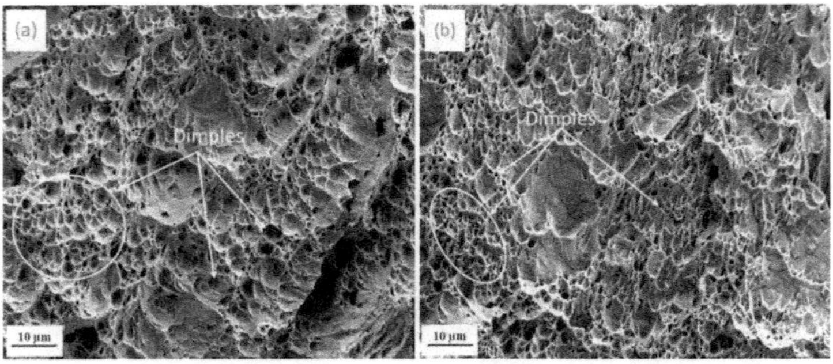

Fig. 4.70 SEM Fractography of Friction Stir Weld AA2219-T87 processed (a) plain shoulder and (b) with spherical ball shoulder square pin profile tool.

4.12 Benchmarking comparison of results

Several studies have investigated the effects of different pin profiles, including square, hexagonal, and octagonal shapes, on FSW performance. The current friction stir welding of AA2219-T87 processed with Square, Hexagonal, Octagonal pin profile with spherical ball shoulder enhancement tool weldments results on Microhardness ,Ultimate tensile strength, Microstructure, Corrosion rate were compared with best industry previous published results. Research indicates that square pin profiles often result in superior weld quality Elangovan, Babu and Balasubramanian (2007,2008,2009). The current research results also yields best results with Square pin profile weldments. almost matching with best below mentioned industrial research publications. The microhardness after friction welding was almost reduced 20-25% with W type chart was observed in the Ju Kang et al.(2016) research work , the current research was almost matching the same pattern with almost same reduction in microhardness.. The FSW after friction welding weld joint efficiency was achieved 72.76% (Malarvizhi.S and V. Balasubramanian (2011) The current research with Square pin profile weldments ultimate tensile strength joint efficiency achieved to 71.81%. the results were very closer to the previous said. Kanwer S. Arora et al (2010) Tensile strength of the welds is significantly affected by welding speed and shoulder diameter and slightly by welding speed. Welding speed is the most significant parameter effecting percentage elongation. Microstructure in nugget consisted of recrystallized grain structure with an average decrease in grain size by a factor of 10. WNZ consists of recrystallized microstructure of very fine grains, formed as a result of the severe plastic deformation and high temperature. "Onion rings" are observed here. In all our square, hexagonal, octagonal pin profile weldments in WNZ onion ring structure formations observed, the ultra-fine grain size reduction was noticed.

The post weld corrosion rate was much reduced in WNZ due to the refinement of finer grains, The current research result also matches the same.

Elangovan and Balasubramanian (2007, 2008) investigated the influence of welding speed and tool pin profile on the formation of the Friction Stir Processing (FSP) zone in AA2219 aluminium alloy. In their study, 6mm thick rolled plates of AA2219 aluminium alloy were cut into 300mm × 150mm specimens for experimentation. The joints were produced using five different tool pin profiles—straight cylindrical, tapered cylindrical, threaded cylindrical, triangular, and square—across three varying welding speeds. A macroscopic examination was carried out to analyze the FSP zone, and the tensile properties of the joints were evaluated in relation to zone formation.

Their findings emphasized that welding parameters and tool pin profiles play a crucial role in determining weld quality. The square pin profile, in particular, was found to enhance material flow, leading to a defect-free stir zone with fine, equiaxed grains, thereby improving mechanical properties. The microstructural characteristics of the Weld Nugget Zone (WNZ), Thermo-Mechanically Affected Zone (TMAZ), and Heat-Affected Zone (HAZ) were comparable to those observed in welds made with square, hexagonal, and octagonal pin profiles.

Additionally, their study highlighted that among the five tool profiles, the square pin profile consistently produced the highest tensile strength, regardless of the welding parameters used. This reinforces the effectiveness of the square pin profile in achieving mechanically strong and defect-free welds, outperforming other tool geometries.

S. Babu et al. (2009) explained that the enhanced properties achieved using tool pins with flat faces, such as square and triangular profiles, are due to their ability to facilitate the movement of plasticized material around the probe. Unlike circular profiles, non-circular pin designs create eccentricity during rotation, which influences the material flow. This eccentricity, linked to the dynamic orbit of the rotating tool, plays a crucial role in the friction stir welding process and must be considered as an inherent characteristic affecting weld quality.

C.S. Paglia and R.G. Buchheit (2008) examined the corrosion behavior of aluminum alloy friction stir welds, highlighting that improving resistance to corrosion and environmental cracking primarily depends on minimizing microstructural sensitization after welding. The overall properties of the weld are dictated by its most vulnerable region, specifically the transition zone between the nugget and the thermomechanically heat-affected zone. In this area, coarsened precipitates significantly influence corrosion behavior. Therefore, the key approach to enhancing corrosion resistance is to either partially dissolve or refine the size of these larger precipitates.

Malarvizhi S. and V. Balasubramanian (2011) conducted a comparative study on the tensile, fatigue, and corrosion behaviour of AA2219 aluminium alloy joints produced using Gas Tungsten Arc Welding (GTAW), Electron Beam Welding (EBW), and Friction Stir Welding (FSW). The base metal, in its original T87 condition, exhibited a hardness of HV 140, which significantly decreased in the weld metal regardless of the welding process. Among the welded joints, the FSW joint demonstrated the highest hardness, measuring HV 110 at the weld center. Although this is lower than the base metal's hardness, it remains the most favourable among the three welding techniques.

In terms of mechanical properties, the unwelded base metal had a yield strength of 390 MPa and a tensile strength of 470 MPa. Among the welded joints, the FSW joint achieved the highest yield strength of 305 MPa and a tensile strength of 342 MPa. The overall joint efficiency was calculated to be approximately 72.76%, which, while lower than that of the base metal, indicates superior performance compared to other welding methods.

Ju Kang et al. (2016) analyzed the microhardness distribution across different regions of an AA2219-T8 friction stir welded (FSW) joint, revealing a characteristic "W" pattern, as shown in Figure 4. The base metal (BM) exhibited the highest hardness at approximately 142 $H_V0.1$, while the lowest hardness value was recorded in the thermo-mechanically affected zone on the retreating side (TMAZ-RS) at around 79 $H_V0.1$. The weld nugget zone (WNZ) had an average hardness of about 90 $H_V0.1$, which was 10 $H_V0.1$ higher than that of the TMAZ. In the heat-affected zone (HAZ), hardness gradually increased with distance from the TMAZ.

Additionally, the study found that the base metal in the AA2219-O condition had significantly lower hardness than that of AA2219-T8. The WNZ was subjected to high temperatures exceeding 748 K (475°C) during welding, leading to the complete dissolution of θ ' and θ' phases and partial dissolution of the θ phase. This dissolution increased the amount of copper in the form of a supersaturated solid solution (SSS). Since SSS is less effective in restricting dislocation motion compared to h¢¢ and h¢ phases, the hardness in the WNZ remained relatively low at around 90 HV0.1. However, the WNZ also underwent dynamic recrystallization during welding, which refined the grain structure and contributed to some degree of strengthening.

Kanwer S. Arora et al. (2010) reported a decline in tensile properties in welded specimens compared to the base material. The tensile strength of the welds was primarily influenced by welding speed and shoulder diameter, with welding speed having a minor effect. Among the parameters, welding speed had the most significant impact on the percentage elongation of the welds. The microstructure of the weld nugget exhibited a recrystallized grain structure, with the average grain size reducing by a factor of 10. In the thermo-mechanically affected zone (TMAZ), microstructural changes were attributed to the combined effects of heat and deformation. The weld nugget, located at the region where the tool pin penetrates the joint, featured a recrystallized microstructure with extremely fine grains. This refinement resulted from intense plastic deformation and exposure to high temperatures during the welding process. Additionally, distinctive "onion ring" patterns were observed in the weld nugget.

CHAPTER-5

CONCLUSIONS

In the current research work, Friction Stir Welding on the AA2219-T87 plates were successfully performed with the square pin, hexagonal pin and octagonal pin profile tools with special spherical steel balls impregnated shoulder enhancement. After successful processing, samples were prepared and exposed to microstructural characterization and mechanical performance evaluation. The following conclusions were derived from the findings of the current investigation.

5.1 Friction Stir Welding bead macrostructure of different tool pin profile welded joints of AA2219-T87

- FSW after-weld bead width varies at the top and bottom according to the different tool pin profiles with special spherical steel balls impregnated shoulder enhancement. The Stir zone looks like a bowl in shape in hexagonal pin profile and an inverted half pot in shape in square and octagonal pin profile welding.
- The upper surface experiences more frictional heat and drastic plastic flow due to the contact with tool's shoulder, which results in a wider upper surface. The bottom portion of the joint material only exposed by the pin profile diameter or size for a shorter width only. It also depends on the tool's linear velocity (travel speed) and rotational speed.
- Among the three-tool pin profile, the square pin profile welded joints have wider bead width with topmost layer was around 19 mm and on bottom most layer was around 8.10 mm., the octagonal pin profile topmost layer was 18.60 mm and at bottom most layer it was around 7.30 mm and the hexagonal pin profile top most layer was 18.70 mm and on bottom most layer it was around 6.10 mm. The octagonal pin profile welded joints have comparatively smaller bead width. The bead width of hexagonal pin profile welded joints is in between the square and octagonal pin profile.
- The average FS bead width varies with tool rotational speed and tool traverse speed. The bead width was wider at 1000 rpm with slow tool traverse speed of 20 mm per min, but higher spindle speed of 1500 rpm with 30 mm per min. tool traverse speed, the bead width was comparatively smaller than the others. The moderate tool rotation spindle speed with slower tool traverse speed makes comparatively higher temperature associated with higher bead width FS weldments

and higher tool rotation spindle speed with comparatively faster tool traverse speed makes narrower bead width in material consolidation was observed.

5.2 Friction Stir Welding bead microstructure of different tool pin profile welded joints of AA2219-T87

- The base metal (BM) exhibits larger, pancaked-shaped grains oriented along the rolling direction, with an average grain size of 63 μm., typically elongated and contain θ ($CuAl_2$) particles distributed throughout the matrix. The EDS analysis indicates that the material primarily consists of α-aluminium (91.75% by weight), with copper as the second major element (6.70% by weight), consistent with the parent metal's chemical composition. The EBSD analysis, specifically the Kernel Average Misorientation map, reveals a lower KAM value, indicating a reduced geometrically necessary dislocation (GND) density. Furthermore, the grain boundary distribution map shows a higher proportion of Low-Angle Grain Boundaries (LAGBs) and a comparatively lower presence of High-Angle Grain Boundaries (HAGBs), suggesting minimal restriction at the grain boundaries.
- The WNZ appears to be different in grain size at the top and bottom due to the difference in thermal heating experience of the top and bottom portions. The top of the nugget zone is exposed to a higher degree of heating and the larger centrifugal force due to the close contact with the tool shoulder makes higher the deformation resulting finer recrystallisation in grain size. The apparent decrease of grain size at bottom is due to the heat sink effect of the steel backing plate, which draws away the heat, making less heat available for grain growth.
- Each rotation of the tool in forward motion with one revolution during FSW causes layers of material that are semi-cylinder shapes to be extruded, known as onion ring structure. The extruding spinning action of square pin, the octagonal pin and hexagonal pin profile tool are varied in grain size.
- Among all the above three tool weldments, the WNZ of the octagonal pin profile grains were ultra fine equiaxed, comparatively smaller in size (average grain size of 3.09 μm) than square and hexagonal pin profile. The square pin profile weldment WNZ grains were comparatively bigger in size (average grain size of 8.86 μm) and the hexagonal profile weldment WNZ grains were in between the square and octagonal pin profile weldments.

- In TMAZ, the parent grains were reoriented along the flow pattern around the tool pin by shear deformation of the rotating tool, In shoulder under it is wider due to extra heat and centrifugal force during stirring and it is narrowest at the bottom as the deformation and heat generated by the tool pin is comparatively smaller. The TMAZ micrographs exhibit coarser grains and so reduced grain boundary strengthening is observed in this region. Thus, the lower hardness of TMAZ is attributed to the low grain boundary strengthening and precipitate hardening. EDX spot analysis of AA2219-T87 TMAZ area further confirms there is a marginal alteration of constituent elements in TMAZ., there is a higher constituent elemental variation in the availability of α aluminium 86.21% by weight followed by next major copper alloy with 12.80 % by weight, The marginal change were noticed, which were different against as base metal constituent elements weight % as against base metal EDX results by the spherical ball shoulder spinning.
- The HAZ just experiences a thermal cycle heating which transferred from adjacent TMAZ and does not undergo appreciable plastic deformation. The temperature was comparatively lower. The average grain size of HAZ for the FSW processed AA2219-T87 was noted as 31.32 μm, the grain size was much larger than the TMAZ.

5.3 Performance of Square pin profile tool with spherical ball shoulder on AA2219-T87 friction stir weldments

- The graphical illustration of square pin profile tool microhardness values appears like "W" in shape. The after weldment highest micro hardness value of 98.63HVwere noticed in the HAZ-AS side of the serial number 2 weldments, the lowest micro hardness value of 83.87 in TMAZ-RS side serial number 2 weldment.
- In the after-weld micro hardness assessment, the FSW method dramatically reduces the hardness of the metal in the weld region, around 22 to 31%. This is due to the significant amount of heat input that causes precipitate disintegration, loss of cold work hardening, and elevated temperatures during FSW. The hardness of the base material is measured in between 117 Hv to 126.92Hv.
- The after-weld transverse tensile properties of square pin profile welded joints was comparatively higher ultimate tensile strength than the other tool pin profile weldments. For the square pin serial number 2, weldments was carried out with the tool spindle rotation speed of 1000 rpm with 3^0 tool tilt angle and 25 mm per min. table

traverse speed yields highest Ultimate Tensile strength of 293 Mpa. with the highest joint efficiency of 71.81%. The comparatively slower tool rotation speed along with a moderate tool tilt angle and with a moderate tool traverse speed influences higher bead width weldment profile material consolidation along with uniformly distributed fine $CuAl_2$ particles joint chemistry with the parent metal.

- On the square pin weld specimen serial number 9, FS welding was performed with the tool spindle speed rotation of 1500 rpm with 3^0 tool tilt angle and 30 mm per min. table traverse speed yields lower Ultimate Tensile strength of 209 Mpa. with joint efficiency of 51.22% .Due to the higher tool rotation speed with faster travel speed with wider tool tilt angle by the comparatively smaller bead width weldment profile material consolidation interrelated micro chemistry.
- From the Potential dynamic corrosion survey results outcome, the square pin serial number 2, friction stir weldments was comparatively lower WNZ Corrosion Rate value of 0.01699 mm per year and Square Pin profile weld specimen serial number 9 has comparatively higher WNZ Corrosion Rate value of 0.04238 mm per year was observed.

5.4 Performance of hexagonal pin profile tool with spherical ball shoulder on AA2219-T87 friction stir weldments

- The graphical illustration of hexagonal pin profile tool microhardness values appears like "U" in shape. The after weldment highest micro hardness value of 101.47H_V was noticed in the HAZ-AS side of serial number 9 weldment and the lowest micro hardness value of 82.70 in the WNZ of serial number 6 weldment. The base metal micro hardness values were in-between 126.92 H_V to 117.73H_V.
- The transverse tensile properties of the weld specimen serial number 2; On FS welding performed with the tool spindle speed rotation of 1000 rpm with 3^0 tool tilt angle and 25 mm per min. table traverse speed and yields Ultimate Tensile strength of 258 Mpa., with joint efficiency of 63.23%, due to the joint no.2 by the slower tool rotation speed with moderate travel speed with moderate tool tilt angle the comparatively broader bead width weld weldment profile material consolidation inter related micro chemistry.
- The weld specimen serial number 9, FS welding was performed with the tool spindle speed rotation of 1500 rpm with 3^0 tool tilt angle and 30 mm per min. table traverse speed and it yields lowest Ultimate Tensile strength of 198 Mpa. with joint efficiency of 48.52%. Due to the higher tool rotation speed with faster travel speed

with wider tool tilt angle by the comparatively smaller bead width weldment profile material consolidation interrelated micro chemistry.

- From the Potential dynamic corrosion survey results outcome, it was observed in the hexagonal pin serial number 2, friction stir weldments have comparatively lower WNZ Corrosion Rate value of 0.027mm per year and hexagonal Pin profile weld specimen serial number 9 have comparatively higher WNZ Corrosion Rate value of 0.04738 mm per year.

5.5 Performance of octagonal pin profile tool with spherical ball shoulder on AA2219-T87 friction stir weldments

- The graphical illustration of octagonal pin profile tool microhardness values appears like "W" in shape. The after weldment highest micro hardness value of 98.82$H_V$were noticed in the HAZ-AS side of serial number 6 weldment, the lowest micro hardness value of 86.08 in WNZ side of serial number 6 weldment. The base metal micro hardness values were inbetween127.62 H_V to 117.91H_V.
- The transverse tensile properties of the octagonal pin profile weld specimen serial number 2: FS welding was performed with the tool spindle speed rotation of 1000 rpm with 3^0 tool tilt angle and 25 mm per min. table traverse speed and it yields higher Ultimate Tensile strength of 256 Mpa. with joint efficiency of 62.7% As per the joint no.2 slower tool rotation speed with moderate tool traverse speed and with moderate tool tilt angle by the higher bead width weldment profile with uniformly distributed fine $CuAl_2$ particles joint chemistry with the parent metal.
- The tensile properties of the octagonal pin profile weld specimen serial number 8: FS welding was performed with the tool spindle speed rotation of 1500 rpm with 2.5^0 tool tilt angle and 25 mm per min. table traverse speed and it yields lower Ultimate Tensile strength of 220 Mpa., with joint efficiency of 53.92%. Due to the higher tool rotation speed with moderate traverse speed and with moderate tool tilt angle associated lower bead width joint micro chemistry of the joint metal.
- From the Potential dynamic corrosion survey results outcome, it was observed that the octagonal pin serial number 2, friction stir weldments have comparatively lower WNZ Corrosion Rate value of 0.03874 mm per year and octagonal Pin profile weld specimen serial number 4 have comparatively higher WNZ Corrosion Rate value of 0.04954 mm per year.

5.6 Ultimate Tensile strength, Microhardness and corrosion rate outputs with FSW Input Parameters through SNN Optimization for various pin profiles

- The proposed SNN-EAO Optimization technique towards the arrival of optimized FSW input parameters in aiming better Ultimate Tensile strength, Microhardness and corrosion rate outputs for the current tool pin Profiles of square, hexagonal and Octagonal spherical ball shoulder tools. The proposed model is evaluated with the existing techniques in terms of performance matrices such as RMSE, MAE, MSE, R-Square, and Correlation coefficient.
- The outputs were predicted based on optimization algorithm with the trained inputs for each property, The Ultimate Tensile strength, Micro hardness and corrosion strength of mid weld of different pin experimental samples were calculated.
- From the experimental result average, actual result average, and the difference between them were calculated. From the average R-values of respective pin profile UTS, MR, CR tables, the RMSE, MAE, MSE, R-Square, and Correlation coefficient were arrived. Based on the SNN outcome optimized input feasible parameters confirmation tests were conducted with the respective square pin, hexagonal pin and octagonal pin profile tools with special spherical steel balls impregnated shoulder.
- For the Square pin profile tool confirmation tests welding, with the optimized feasible FSW input parameters of Tool rotation spindle speed of 1100 rpm , 25 mm per min table traverse speed and 3^0 tool tilt angle , the confirmation test yielded an UTS of 291.70 MPa, a MH of 97.6 Hv, and a CR of 0.0169 mm/year. The percentage errors for TS, MH, and CR were 3.45633%, 2.7652%, and 1.8221%, respectively. The R-value was 97. 3187.
- For the hexagonal pin profile tool confirmation tests welding, with the optimized feasible FSW input parameters of Tool rotation spindle speed of 1100 rpm, 25 mm per min table traverse speed and 3^0 tool tilt angle, as per the confirmation test yielded an UTS of 258 MPa, a MH of 82.97 HV, and a CR of 0.03465 mm/year. The percentage errors for TS, MH, and corrosion strength were 1.24396%, 1.2849%, and 13.375%, respectively. The R-value was 94. 6986.
- For the Octagonal pin profile tool confirmation tests welding, with the optimized feasible FSW input parameters of Tool rotation spindle speed of 1250 rpm, 25 mm per min table traverse speed and 3^0 tool tilt angle, the confirmation test resulted in

a TS of 234 MPa, a MH of 93.14Hv, and a CR of 0.04891 mm/year. The percentage errors for TS, MH, and CR were 3.4613%, 7.41552%, and 2.18%, respectively. The R-value was 95.6477.

- Based on the outcome of confirmation test results demonstrate the respective pin profile output properties, based on average R values, for all experimental and confirmation test results, the outcomes were within very small margins of error. The confirmation test results were matching with actual test and confirms around 95% confidence level.

5.7 Comparative study of square pin plain shoulder tool FS weldments with square pin Spherical ball shoulder tool FS weldments

- The hardness value in WNZ of the specimens produced with spherical ball shoulder was comparatively higher than the hardness values of the specimens produced without spherical ball shoulder. AA2219-T87 alloy processed with spherical ball shoulder was increased than the sample processed without spherical ball shoulder may be due to the comparatively lower frictional heating of shoulder having spherical balls.
- The grain size of the WNZ for the AA2219-T87 samples processed with spherical ball shoulder was comparatively smaller than the samples processed without spherical ball shoulder. The size of the grains was significantly affected by the spinning effect of spherical ball shoulder on the WNZ, TMAZ, HAZ.
- The tensile behaviour of the AA2219-T87 samples processed without and with spherical ball shoulder. The yield and ultimate tensile strength of AA2219-T87 specimens processed with spherical ball shoulder were comparatively higher than the samples produced with plain shoulder. The YS and UTS of the AA2219-T87 specimens processed with spherical ball shoulder were improved by 11.17% and 7.39% than the samples processed without spherical ball shoulder. On the other hand, the percentage elongation was found reduced with spherical ball shoulder FS weldment tensile sample than the plain shoulder weldment sample.
- EDX confirms the changes in alloy composition across the weld zones of AA2219-T87, the elemental concentration measurement analysis by SEM morphology reveals there is a thermal influence and frictional heating impacts from the tool shoulder spherical ball with the pin typically altering the constituents and grain size variations in the WNZ, TMAZ. On WNZ, the EDX spot analysis reveals there is a huge variation in the availability of α aluminium of 59.10% by weight followed by

major copper alloy with 23.50 % by weight. In TMAZ area there is a higher constituent elemental variation in the availability of α aluminium 86.21% by weight followed by major copper alloy with 12.80 % by weight. (AA2219-T87 Base Metal, EDX reveals the availability of α aluminium of 91.75% by weight followed by major copper alloy with 6.70 % by weight)

- From the after tensile samples SEM fractography there, is a greater number of deep and wider dimples were noticed in the FS Weldments of AA2219-T87 samples processed with plain shoulder, and on the other hand fine dimples were noticed in the samples processed with spherical ball shoulder.

5.8 Identification of best FSW tool pin profile based on the experimental outcome results

- The after-weld Ultimate Tensile Strength of square, hexagonal and octagonal pin profile tool weldments was compared and the results were as follows. The higher Tensile strength value was achieved with hexagonal pin profile welded with test coupon no:2 with a value of 258 Mpa with weld joint efficiency of 63.23%. The higher Tensile strength value was achieved with octagonal pin profile welded with test coupon no:2 with a value of 256 Mpa. with weld joint efficiency of 62.74%. The highest Tensile strength value was achieved with square pin profile welded with test coupon no:2 with a value of of 293 Mpa. with weld joint efficiency of 71.81%. The square pin UTS having the highest value than the hexagonal and octagonal pin weldments. Optimal welding performance was attained with a spindle speed of 1000 RPM, a travel speed of 25 mm/min, and a tilt angle of 3 degrees.

- The after-weld micro hardness assessment, the FSW method dramatically reduces the hardness of the metal in the weld region, comparing the WNZ microhardness of square pin, hexagonal pin and octagonal pin weldments. The WNZ of square pin weldments having highest microhardness value of 94.6H_V in the test coupon no:2. The WNZ of hexagonal pin weldments having highest microhardness value of 84.9H_V in the test coupon no:7. The WNZ of octagonal pin weldments having highest microhardness value of 93.14H_V in the test coupon no:5. The square pin WNZ having the highest micro hardness value than the hexagonal and octagonal pin weldments.

- ❖ From the Corrosion rate in mm /year of WNZ of square, hexagonal and octagonal pin profile tool weldments, The highest corrosion strength value of 0.01699 mm/year was achieved with Square Pin profile tool, subsequently the next high corrosion strength value was achieved with a Hexagonal Pin with strength value 0.03165 mm/year and finally with Octagonal Pin 0.03874mm/year. The square pin profile yielded the highest corrosion strength than the hexagonal and octagonal pin profile tool weldments
- ❖ Based on the current research work with the outcome results on microhardness, ultimate tensile strength and corrosion rate results of the respective square pin, hexagonal pin and octagonal pin weldments, The square pin profile with spherical ball shoulder tool weldments out performed than the hexagonal and octagonal pin profile tool weldments. The square pin profile with spherical ball shoulder tool can be the better choice for further practical applications

5.9 Scope for future work

- ➢ The Spherical ball shoulder enhancement tools of square, hexagonal and octagonal pin profiles FSW weldments outcome properties like UTS microhardness, corrosion properties were checked further Creep, Torsional, Wear properties can be carried out and analyzed
- ➢ After post weld heat treatment tensile, shear, compressive properties can be checked
- ➢ Triangular and pentagon tool pin profile tool with spherical ball shoulder tool welding and further checking of mechanical, corrosion, wear and creep properties may be checked
- ➢ Long-term Mechanical performance and durability of welded joints under cyclic loading and environmental conditions to determine their suitability for various industrial applications.
- ➢ Advanced Characterization Techniques such as transmission electron microscopy (TEM) and atom probe tomography (APT) to gain deeper insights into microstructural changes

www.ingramcontent.com/pod-product-compliance
Lightning Source LLC
LaVergne TN
LVHW010722061126
851857LV00014B/1302